Perpetual
Motion

by the same author

COLLECTING MUSICAL BOXES
PLAYER PIANO
MECHANICAL MUSIC
CLOCKWORK MUSIC
BARREL ORGAN

PERPETUAL MOTION

The History of an Obsession

ARTHUR W. J. G. ORD-HUME

London

GEORGE ALLEN & UNWIN LTD

RUSKIN HOUSE   MUSEUM STREET

First published in 1977

© George Allen & Unwin (Publishers) Ltd 1977

ISBN 0 04 621024 5

Printed in Great Britain
by Butler & Tanner Ltd
Frome and London

ACKNOWLEDGEMENTS

Compiling this book has been an extremely interesting task and to all the friends, associates and correspondents who have given so freely of their help and knowledge I express my sincere thanks.

My special thanks to the Franklin Institute in Philadelphia for allowing me access to a great deal of perpetual motion material. Particularly I acknowledge the help received from Albert D. Hollingsworth, director of public relations, Charles Penniman, assistant director of educational operation, Walter Pertuch, emeritus librarian, Emerson Hilker, chief librarian, and Mrs Linda Wohlforth, the reference librarian.

The US Patent Office Library at Arlington County, Virginia, and the office of the Commissioner of Patents, Washington jointly produced much about the American perpetual motion inventors. The British Museum, the library and collections of the Guildhall, the Victoria & Albert Museum and the Science Museum, South Kensington, have provided help and access to reference material. For their help and advice, I acknowledge the assistance provided by the late Dr Horace Manley, B.Sc., Ph.D., and G. R. Clarke, B.Sc., of the United Kingdom Atomic Energy Authority.

Douglas Berryman, owner of the West Cornwall Museum of Mechanical Music, near Penzance, has provided me with much reference material from his own library. Thanks also to Dr Jacob Nevyas, Emeritus Professor, Pennsylvania College of Optometry, Philadelphia who, in wishing to be recorded as a staunch disbeliever in anything to do with perpetual motion, has proved invaluable in locating material for me and guiding me through library, museum and records collections in the United States. Dr Howard Fitch of Summit, New Jersey has kindly located and copied some of the less-known documents for me.

I must not omit my indebtedness to those writers who have chosen to commit to paper aspects of this subject before me. As regards periodicals (and I rate these highly for it is they who record the more ephemeral happenings in the world which get forgotten or otherwise cast aside when more permanent works are compiled), I have made extensive use of material from the *Scientific American*, the *English Mechanic, Mechanics, Nature* and the various digest magazines such as *Cassier's, The Strand* and others. I have quoted from the works of the meticulous Henry Dircks and, of more recent origin, Clifford Hicks of *Popular Mechanics*, Dr Henry Morton, a former president of the Steven Institute (1894–5), Professor Coleman Sellers (1895), Professor Hele Shaw of University College, London (1887–8), F. Charlesworth, assistant examiner in the British Patent Office (1905), Charles E. Benham (1916), Dr Paul R. Heyl, US Bureau of Standards (1926), Wilson C. Morris (1913), John Walker Harrington (1929), William Marion Miller (1947), Dr G. Kasten Tallmadge (1941), Silvio A. Bedini (1956), and Stanley W. Angrist (1968). The dates in parentheses refer to works detailed in the Bibliography.

For sowing a very early seed of interest in so many and varied things, my late father deserves my humble gratitude. Although neither misguided nor an inventor, he was intrigued by this unattainable, probably more from the philosophical, almost poetic viewpoint than anything else. He used to tell me about it all when I was very small. Whilst I was incapable of true understanding, the words 'perpetual motion' etched their way into the corner of my burgeoning mind, so this book is some sort of a memorial to his ability to open and intrigue a young mind whereas so many today are never given such an opportunity.

Some measure of thanks should also go, ruefully, to those responsible for my mechanical and engineering education, particularly to one university professor who, in response to my numerous queries as to why perpetual motion could not be achieved by various means, ultimately forbade me ever to mention the subject again. He left to join the Great Majority many years ago, but if he should get to hear of this book, I beg his forgiveness.

In conclusion, I owe thanks to the BBC and to Karl Sabbagh, TV producer, who first had the idea of doing a *Horizon* programme on perpetual motion. It was largely through my researches for that programme that a long-standing file of old references and material ended up in the form which you now have in your hands.

PREFACE

This book started in an unusual way. During my researches into the history of automata and mechanical musical instruments, some of the results of which have already been published in my other books, I kept coming across references to perpetual motion. These references were too numerous and too intriguing and my own mind too inquiring for me to ignore them. So I began a file for references to perpetual motion. I confess that I felt somewhat self-conscious about the very existence of such a file, particularly since my early training in engineering and the science of mechanics had taught me that there could be no such thing as *perpetuum mobile*. However, the file continued to grow at such a rate that my curiosity got the better of me and I started to take more than just a passing interest in the subject of its contents.

A catalyst to my interest came when the BBC directed to me a plea for help in research for a television programme on perpetual motion a year or two back. For several weeks I became thoroughly steeped in this subject, exploring in depth the references which I had filed away over the years.

What I found was remarkable, exciting and often very amusing. I found several mysterious devices which, rather like that small percentage of Unidentified Flying Object sightings, cannot easily be explained away. I found that whilst the majority of perpetual motion seekers were lacking in a fundamental understanding of the principles involved, some, amongst whom are famous names in the sciences, devoutly believed in perpetual motion not just as a possibility but as a means of providing power to drive Man's machines. There is a tragic side to the story. Tales of men of great ability who devoted their whole lives to seeking perpetual motion, who went mad as a result, who squandered fortunes only to die penniless, broken and unfulfilled men. Many faced such a sad anti-climax to their quest. Other men were more cunning than deceived in their ideals. They resorted to deception and some saw the quest for discovering perpetual motion as a means toward ill-gotten gains: today we would call them confidence tricksters. Strangely enough, many of those who resorted to deceit were driven to such a course by the failure of what set out as a genuine attempt to find a self-acting force.

Before embarking on preparing a book on this subject, I took great pains to set out the exact approach which I would make to my subject. Should one narrate fact without comment and produce a sterile obituary—a testimony and nothing else? Or should I try to steer my readers through a subjective history of the quest for the impossible, explaining why the various attempts failed? I read the eminent nineteenth-century work of Henry Dircks, which falls into the first category, and settled for the second.

Armed with the practical dogmas of an engineer, the philosophical approach of a thinker and the inquiring mind of a life-long student, I shall attempt to discharge my self-imposed task in the best possible manner. I have also allowed myself to depart from the rigid doctrines of the historian and look just a little further afield to evaluate whether or not man-made perpetual motion is now already with us.

I would add one word of caution. Judge a subject by those who have chosen to work at it. If the calibre of those men is such that we might respect their judgement, then we must equally respect their work, however misguided it may, with the wisdom of hindsight, appear to have been. Perpetual motion is a serious subject. It is not the overtly stupid thing which some inventors have presented to the public. Like the alchemist, the perpetual motion seeker was a dedicated scientist, fettered not by his own ignorance, but by the limitations of scientific and mechanical knowledge which pertained at the time. These men, in the main, sought to extend science in a manner which to them seemed natural at a time when Man's knowledge and understanding about so many things was advancing at an evocative pace. It should be to the everlasting satisfaction of men such as these that their dreams have in some ways come true and that alchemy and perpetual motion are in a sense twentieth-century stated facts.

It might also be prudent to remember that over the years there have been many noted men of letters whose abilities have been fettered by the surprising narrowness of their intellect. When first it was proposed to oust the sail in favour of a steamship to cross the oceans, no less a person than the once-celebrated Dr Lardner loudly proclaimed, 'No steam-ship could be built large enough to carry sufficient coals for a voyage across the Atlantic!'

The history of the world is peppered with historic and monumental failures, almost all of which might have been averted if adequate thought by sufficient intellect had first been applied.

The perpetual motionists by their very failure advanced our knowledge and our scientific understanding, and many were the things discovered thanks to what today we would call 'technological spin-off'. The 1939–45 war in Europe advanced by twenty years the technology of the aeroplane in the space of but five, and the Vietnam war benefited the development of the helicopter by more than a decade, so demonstrating that a measure of good may be generated by evil. Atomic science might not have reached its present state had it not been for the terrible deaths of Hiroshima and Nagasaki. And so, in a much more mundane way, the world benefited from this quaint facet of experimental science—the search for automatic energy.

And if there still be doubt as to the wisdom and far-sightedness of some of our able forefathers at least, let me quote from the writings of Charles Louis de Secondat, the Baron Montesquieu (1689–1755). Political theorist, philosopher and author of *Persian Letters* published prior to the year 1721, he showed great sagacity and perception when he wrote:

'Everywhere I see people who talk continually about themselves ... Some days ago a man of this character wore us down for two hours ... but as there is no such thing as *perpetual motion* in this world, he stopped talking.'

Arthur W. J. G. Ord-Hume

CONTENTS

INTRODUCTION

Just as soon as Man discovered the processes of elementary mechanics, he became interested in the possibility of having his machines work for him by themselves. To these early artisans with their simple prime needs, automatic machinery was seen not so much as a goal to be achieved as just another aspect of the work of the wheelwright and blacksmith. Perpetual motion surrounded him on a grand scale—the sun rose and set, the moon waxed and waned, the seasons changed, the tides ebbed and flowed. Water and wind cost nothing and their natural abundance could not be overlooked. The forces of Nature existed everywhere and their harnessing was itself seen as a form of perpetual motion. What was needed to grind corn automatically, or pump water continually was considered nothing extraordinary and the ways of having your work done for you were believed many and varied.

Hence it is not surprising to discover that all the early perpetual motion machines concerned the craft of the miller who used either water or wind to grind his corn. He only had to site his waterwheel in a steadily flowing stream of water and he had all the power he needed without recourse to horses, capstans and wheels. The waterwheel could turn heavy millstones one upon the other and could even drive a sack-lift so that the miller could be spared the humping of bags of grain or the finished flour. Until, that is, the dry season or any other event which served to reduce the flow of water in the stream.

Now the mathematician Archimedes, born 287 BC, showed how water could be lifted by using a screw in a tube. This at once made life absurdly simple. All the miller had to do was to drive an Archimedean screw and once the water had passed through the waterwheel, lift the water back up to the top and let it do its work all over again. The screw, needless to add, could be driven by the waterwheel—and the whole process would become continuous. Why, you did not even need to have a stream of water, just a pond with a fixed volume of water which could be raised and lowered *ad libitum*.

Happily for the millers, their water supplies must never have dried up, for if this had happened it would have become embarrassingly obvious that some serious defect prevented the otherwise carefully worked out system from working.

Perpetual motion, it seemed, was always close, yet the ineluctable laws of motion and energy conservation, of which they could then have had no knowledge, systematically denied them success.

This, then, was the typical example of the true perpetual motion seeker's endeavours—honest, simple-minded, practical in his needs and all out to make a thoroughly utilitarian machine for his own use or that of his family.

There were other kinds of perpetual motion inventor such as those who worked hard along an ill-advised path of self-destruction. Some made machines which all but worked, and then sought (and sometimes gained) financial support for the work claimed necessary for the completion of their machinery—a task which, of course, was never accomplished. They were prevented by self-deception from realising that to gain that few final per cent of efficiency which separated failure from success was an impossibility.

Then there were qualified and respected scientists, mechanicians and engineers who devoted their time to seeking what was impossible. Richard (later Sir Richard) Arkwright, inventor of the spinning jenny, sought the chimerical power. So did George Stephenson, perfector of the steam locomotive. Of the others, more or less nondescript, one was reputed to have gone mad, others committed suicide and many underwent changes of character as a result of their unfulfilled dreams. Among these were the men who devoutly believed that they owed it to the world to discover the secret of perpetual motion, and that they alone must create a device that would do work for nothing. Fortunes were freely spent in attempting that goal by those who were quite oblivious to the hard truth that success lay aeons beyond their reach.

By comparison, there were the charlatans. These fell into two classes. There were those who strove hard to make a perpetual motion machine and then, finding their labours frustrated, employed some form of trickery to make out to the public at large that they had succeeded. Often their final modifications to their models showed extreme skill in execution, concealed clockwork being added in an ingenious way. The other class of charlatan designed his perpetual motion machine as a fraud from the word go, and relied on the gullibility of the masses (for whom perpetual motion was seen as a panacea within the grasp of a clever man) to enable him to extract money from investors. He, too, was usually a clever engineer and perhaps, in reality, he was the cleverest of all the perpetual motion seekers in that he knew from the start that his professed quest was not worth embarking on and so set out with the intention of achieving by deception what the more foolhardy sought to attain by fruitless endeavour.

Perhaps the saddest group were those ordinary people who clamoured to invest their life's savings, the industrialists, financiers and even politicians who fervently believed that they were in on the discovery which would make history and give them untold wealth. There was the Keely motor, but the classic example must surely be the events in the United States in the early part of this century when a perpetual motion machine nearly made Congress a laughing stock after the President had appointed a team of experts to look into it.

Today we may laugh at the credibility, even stupidity, of our forefathers for even countenancing such far-fetched proposals. *Tempora mutantur, nos et mutamur in illis!* The only real thing that we must criticise our predecessors for is their inability to reason that perpetual motion in the manner after which they sought it was quite impossible. Had they possessed the broadness of intellect to encompass that which is a scientific and mechanical commonplace today, then things might have been different.

And those who did use sheer genius to discover a way towards perpetual motion had not the farsightedness to realise that to build something which is intended to go 'forever' you have to select very carefully the place where you build it. James Cox's perpetual motion clock might have been in motion to this day had it not been for the fact that he built it so that it could not be moved—and his premises were demolished something like one-and-a-half centuries ago.

The age in which we live is a harshly materialistic one and, taken out of context, the history of this once-named science must sound inconsequential and vaguely comic. Pause, though, for a moment and visualise simpler times before even the age of steam, when power meant wind, water or the horse, and the world was still in the Dark Ages. Obliterate the last two hundred years of knowledge, of learning, of understanding. Now you may see, in the dimness, how tangible perpetual motion seemed and, perhaps, you may understand how simple and yet important was the quest for *perpetual mobile*.

1

What is Perpetual Motion?

Find something that does more work than the energy you put into it—and you have solved perpetual motion!

A perpetual motion machine, in order to work, must create energy. It must pluck from the ether such a plethora of energy that it can operate without being provided with identifiable external power.

If one were to set up a cash-style balance-sheet for energy relating to any operation you care to name, from pushing a lawnmower to driving an ocean liner, to hammering in a nail or flying at twice the speed of sound, the total amount of energy being supplied (the left-hand column of our balance-sheet) must always equal the amount of energy produced or emanating from the operation—the right-hand column.

The energy which we may loosely speak of as 'lost' has not in truth disappeared. It has merely changed its form so that it ceases to become possible to convert it into mechanical or electrical energy. The losses which occur are due to heat, usually by friction. This is, generally speaking, true of all energy losses for they nearly always transform ultimately into heat.

This can be expressed in another, less complicated manner. In all operations, such as those listed above, the sum total of energy at the end of the operation is the same as it stood at the beginning. It has probably changed its form; it may even have become of little or no use. The heat produced in an internal combustion engine, for example, is a by-product of energy conversion; it is not a necessary by-product, but an unavoidable one. We may make use of some of it to heat our car, but whether we do or not, part of the work done by the car engine is lost to heat. This is expressed as the principle of conservation of energy.

For the perpetual motion machine, though, it must continue to do external work without being supplied with energy. In simple terms, no driving effort must be applied, and no fuel burned.

There is some justification for the statement that it was this very quest for the impossible which laid the foundations of our knowledge of mechanics.

Wise men, the founders of our modern scientific knowledge, among whose ranks there numbered those whom today we consider to be our greatest early scientists, adopted the assumption that perpetual motion was an impossibility, and from this assumed impossibility they discovered the key to the burgeoning science of mechanics.

Whilst it may be easy to prove that a given perpetual motion scheme will not work, and thereby deduce that perpetual motion if sought by that particular means is an impossibility, it is not always so easy to prove that perpetual motion must automatically be impossible by any means. It is for this reason that the impossibility of perpetual motion, by mechanical means, is but an assumption. It is a matter of observation and deduction. It remains impossible to consider the tenet of perpetual motion within the framework of the law of the conservation of energy. Unlike the assumed and therefore probable impossibility of perpetual motion, though, the law of the conservation of energy is one which can be proved and thereby justified from many standpoints.

The man who quantified the laws of motion was Sir Isaac Newton (1642–1727). He found that motion could be classified under three headings—he called them Laws. It is as well to set these out here and now.

First Law   A body once at rest, will remain at rest, unless acted upon by some external force; and if moving in any direction, will continue to move in that direction, unless acted on by some external force.

Second Law   When a force acts upon a body in motion, the effect of this action is the same, in magnitude and direction, as if it acted on the body at rest.

Third Law   Action and reaction are equal and contrary.

In spite of Newton and his undeniable laws, Man has made mechanisms which have sought to be perpetual, and, by some interpretations, they have succeeded. The greatest problem, though, turns out to be not so much Sir Isaac Newton and his Laws as the definition of the word 'perpetual'.

Our understanding of anything is dependent on definition. Even those things which are generally considered to be intangible such as the emotions, sensations of pain or pleasure, sight and sound have, if not exactly a finite form, a form sufficiently capable of description as to enable understanding and identification by others.

When we talk of a dimension, it is pointless to refer to the type of dimension unless we qualify that by a recognisable incremental scale. For example, I cannot communicate the size of an object without providing two pieces of information: the unit of measurement, and the correct number of those units which go to make up the distance I wish to talk about. Two inches, a third of a yard, five miles—we know just what is being described. Even if we are unfamiliar with the units themselves, their meaning can be implied. So six kilometres, twenty-five hectares, a dozen leagues or five degrees of latitude all produce some sort of a picture.

The same goes for colour. I may say that a flower has colour, but that only conveys a general impression. Although I can talk in broad terms of a blaze of colour, I must define that colour if my message is to get across to another party. Prices, state of health, temperature—all are in effect scales against which we use our judgement and experience to interpret.

But what of perpetual motion? What does 'perpetual' mean? How does one define perpetual? Ten years? A hundred, perhaps? Maybe a thousand or a million?

The immediate reaction on meeting the term for the first time may be to interpret it literally as meaning moving 'forever'. But persons of varying training and intellect may assign different meanings. The student of physics will say that a perpetual motion machine is one which does useful work without drawing on an external source. In other words, it is a machine in which the output is greater than the input. The ultimate definition is that here is a machine which will create energy. In recent times, it has been proved that the second law of thermodynamics (this is dealt with in Chapter 2) may not be completely true under all conditions. Still, though, we stumble over 'perpetual' and its meaning.

Our difficulty is complicated the more when we consider that we tend to evaluate 'a long time' in terms of our own time scale—

the statutory, legendary three-score-and-ten years. What, then, of the giant turtle which lives and breathes in our own environment and can live to be 350 years old? Whilst seventy years may be a long time to us, it is only a fifth of the life-span of the turtle. At the other end of the scale, certain species of insect are born with the rise of the sun and live out their tiny lives between then and sunset to die within half a day. To the mayfly, then, four hours is equal to thirty-five years of Man, 175 years of the turtle, and maybe a mere 1,500 years of the life-span of the giant Californian redwood tree.

For something to be 'perpetual', we may naturally consider the word perpetual to be synonymous with 'forever'. In astronomical terms, even 'forever' just isn't good enough. The life of the sun cannot be considered as forever, and the expanding Universe cannot in truth continue for *ever*. Without going into the depths of astronomy, the universe is an odd example of perpetuity. Its very nature is perplexing to the extreme, and this is considered by the astronomer to be one of its most endearing features. There is considerable difference between accepting a situation and understanding it. With the discovery of quasars, in particular one discovered at the Steward Observatory, Arizona, in 1973 and given the euphonious code name OH471, Man's exploration of the Universe has broadened, but his understanding of it has been confronted by yet another mystery. This particular quasar or radio star is on the fringe of the universe and, at the time of writing, it is the most distant object known. It is travelling away from us at a staggering 177,000 miles a second, a speed comparatively easily determined by the Doppler effect which cuts the spectrum of its light as received by us on earth down to a deep red colour. OH471 is receding from us at this incredible speed—90 per cent of the speed of light—and has been doing so for aeons, and will continue to do so, presumably, for eternity.

Is this perpetual motion? Rationally speaking, yes, but then so is the motion of our own solar system around our sun, and the motion of all that is within it—the moon, the earth, our tides, and the winds.

Yet if we look deeper into this, we realise that the motion of the Universe conforms with the precepts of the conservation of energy and Newton's second law, only the very grand scale of the operation serving to blind us, momentarily, to the fact that energy is being used on an equally massive scale to bring about the interaction of all the galaxies within the boundless chamber of space. For here we have something which is probably easiest to comprehend if we describe it as a multi-orientated linear motor with energy in abundance.

Since this energy and the particles of matter which make up that which we loosely call *space* function in a manner which appears effortless and intangible, a form of perpetual motion is assumed to exist external to the earth.

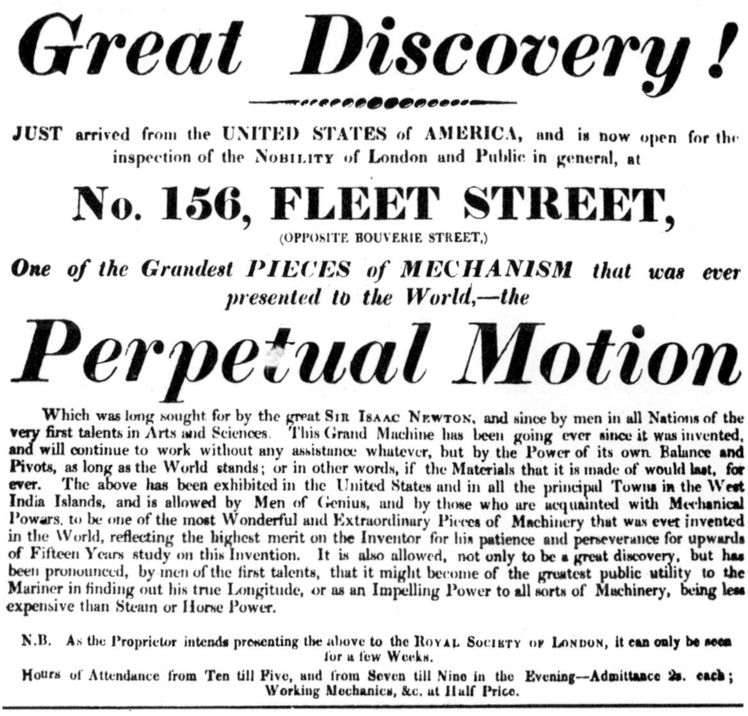

Great Discovery !

JUST arrived from the UNITED STATES of AMERICA, and is now open for the inspection of the NOBILITY of London and Public in general, at

No. 156, FLEET STREET,
(OPPOSITE BOUVERIE STREET,)

One of the Grandest PIECES of MECHANISM that was ever presented to the World,—the

Perpetual Motion

Which was long sought for by the great SIR ISAAC NEWTON, and since by men in all Nations of the very first talents in Arts and Sciences. This Grand Machine has been going ever since it was invented, and will continue to work without any assistance whatever, but by the Power of its own Balance and Pivots, as long as the World stands; or in other words, if the Materials that it is made of would last, for ever. The above has been exhibited in the United States and in all the principal Towns in the West India Islands, and is allowed by Men of Genius, and by those who are acquainted with Mechanical Powers, to be one of the most Wonderful and Extraordinary Pieces of Machinery that was ever invented in the World, reflecting the highest merit on the Inventor for his patience and perseverance for upwards of Fifteen Years study on this Invention. It is also allowed, not only to be a great discovery, but has been pronounced, by men of the first talents, that it might become of the greatest public utility to the Mariner in finding out his true Longitude, or as an Impelling Power to all sorts of Machinery, being less expensive than Steam or Horse Power.

N.B. As the Proprietor intends presenting the above to the ROYAL SOCIETY OF LONDON, it can only be seen for a few Weeks.

Hours of Attendance from Ten till Five, and from Seven till Nine in the Evening—Admittance 2s. each; Working Mechanics, &c. at Half Price.

Fig. 1. During the latter half of the eighteenth and the first half of the nineteenth centuries, London was the showplace of inventions of all types. The notice above dates from the opening decades of the last century and illustrates the compelling style of advertising of the period.

An interesting theory, postulated by the author some years ago as a focal point for discussion, was based on spatial relativity. This is explained with a simple example. If you are expecting motion and something else is seen to move, it is a quirk of the human senses to misappropriate the motion as of oneself. This can be demonstrated in a train. Often when your train is stationary in a station, and another draws alongside, the illusion is that you must be moving and the other train is stationary. So realistic is this that in some instances your own train may start to move as the other slows, giving the impression of relative motion with a stationary object. The suggestion was put forward that the radio stars—the quasars—in the Universe are, in fact, vast stationary powerhouses and it is the entire remainder of the Universe which is speeding in a regulated formation past these in the manner of a sailing regatta negotiating marker buoys. This postulation, which is not as wild as, say, the flat earth theory, cannot, incidentally, be successfully disproved on scientific terms, but breaks down when analysed in pure terms of logic. It also plays some amusing tricks on the perpetual motion concept of the Universe.

But to return to comprehensible parameters in order to define what we mean by perpetual motion, we need to come much nearer to home to determine an acceptable definition. We need to see perpetual motion in the steady state, for motion in the Universe is of little value to Man. Perpetual motion has to be used to do something.

The *Concise Oxford Dictionary*, in telling us that it comes from the Latin *perpetualis*, defines 'perpetual' as meaning:

'Eternal; permanent during life; applicable, valid for ever or for indefinite time; ...'

It goes on to describe perpetual motion in the following terms:

'p. motion (of machine that should go on for ever unless stopped by external force or worn out); continuous; ...'

It becomes apparent that even the compilers of the dictionary are not certain just what the word perpetual really means, for in the above references we have two distinct interpretations—eternal being the first, and permanent during life being the second. These are in effect repeated in the subsequent definition: for ever or for an indefinite time.

Regarding the perpetual motion machine, we find it described as a device which shall go on for ever unless ... (until) ... worn out.

Here, then, we see that perpetual is neither a finite term nor an infinite one. It falls, confusingly, between the two, with the stronger inference that it is indefinite. That this is no doubt necessary is the rationale that perpetuity is indemonstrable.

There is one moment in time when the law of inevitability dictates that the end will come to anything and everything—perpetuity will finally cease. It is only the acceptance of the extreme odds against that moment occurring during our association with a place, an object or event that gives us confidence to exist. Those for whom the odds appear unrealistically short are usually diagnosed as being mentally disturbed. They are not, however, to be confused with those who, through reason, knowledge, experience or perception, or any combination of these qualities, justifiably fear disaster in a specific instance. The discovery of a faulty master cylinder on his car may well suffice to cause its intelligent owner to fear imminent brake failure—a latent disaster—and to take such steps to avoid it by not driving.

The inevitability of the end is apparent, generally speaking, to a lesser extent to the young than to the old, but this is accountable for, usually, by the sharpened perceptions gained by the older through wisdom.

Although the ultimate end of our planet may, in an astronomical sense, be sudden, the end may not be such a terminal holocaust as the proclamations of the religious poster-purveyors might have us believe. In the connections which concern us here, the end is a *serial* business—it is more usually a continuing process. We see the redevelopment of building sites with the tearing down of an old block and the erection of something new which, in its turn, will eventually be demolished. Man himself is a clear-cut illustration of the continuing end since his allotted three-score-and-ten is terminated by death.

In our quest for perpetual motion, then, we may find that the only natural thing which is *perpetual* is this serial existence of life itself.

The philosophers and the engineers considered the quest for perpetual motion from two somewhat opposed standpoints. The philosopher saw perpetual motion as a means of extending a man's prowess and skills into perpetuity and believed that to make something 'perpetual', be it motion or monument, was an achievement concomitant with status, standing or rank. The many statues of forgotten noblemen, gentry and soldiers which formerly stood silent watch over so many of our city squares could be considered, if not truly perpetual motion, then a form of perpetual existence. The Egyptians succeeded in achieving an admirable degree of perpetuity in a similar manner with the pyramids since, unlike the many now-fragmented temples created artistically, these were devised as the ideal shape for both durability and distinction. The tombstone is a similarly contrived, if less ostentatious, system of perpetuity.

Of the philosophers who sought 'real' perpetual motion, their efforts were directed towards the artistic. Great wheels with over-balancing weights combined aesthetic appeal with visual interest. It must be to our everlasting regret that the modern computer was not designed by such a philosopher who would certainly have given it visual attraction.

Now the engineers saw perpetual motion in a different light altogether. Rather than just have something moving for the sake of motion, they sought to make their appliances serve Man: the perpetual motion machine became an engine—a prime mover. Not only had the perpetual mechanism to be capable of overcoming the laws of physics; it had to overcome them with sufficient margin in hand to serve as an unmistakable source of power to drive the wheels of industry.

None of them did, though.

2

Elementary Physics and Perpetual Motion

In the previous chapter I made mention of the laws of thermo-dynamics, and stated that the second law is not rigorously true under all conditions. I also said that mechanical perpetual motion was a myth. Just what these natural laws which shatter the perpetual motion question really are I shall now attempt to explain.

Until the introduction of the concept of energy and the formulation of the principle governing its conservation, it was generally accepted that perpetual motion was impossible. However, this belief rested not upon any general principle but on the evidence that a careful examination of all the proposed devices claiming perpetual motion revealed that each possessed some theoretical flaw which would without question literally put the brake on such a device working of itself.

Sages, mathematicians and engineers alike played their parts in exploding the accepted criteria of perpetual motion, stating that the creation of energy was impossible. Once it had become established, the principle of the conservation of energy became a veritable impenetrable door against which all subsequent claimants to the invention of perpetual motion found themselves knocking in vain.

Soon afterwards, however, the statement of another general classification was evolved. This became known as the Second Law of thermodynamics and it stated, in simple terms, that heat could not run uphill by itself; that if a hot body was placed in contact with a colder one, the tendency would be to equalise their temperatures, not to increase the difference.

No specific reason was first assigned to this tendency. As originally formulated by the German physicist Rudolph Julius Emanuel Clausius (1822–88), the Second Law was purely empirical. Although it was analogous that water would flow downhill under the influence of gravity, the case of the flow of heat was not so easy since there was no recognisable controlling force. Generally, though, heat did follow this downhill path, but to state that it *always* did raised doubts in the minds of many who sought proof of either the irrefutability of the theorem, or its invalidity. By the last quarter of the last century, many notable scientists expressed their doubts as to the generality of the Second Law of thermodynamics and some at least sought experimental proof that a case existed where the law would break down.

In 1875, Maxwell published his then famous *Theory of Heat* in which he postulated that the operation of the Second Law could be suspended by the interposition of intelligence. Maxwell foresaw the great strides which were to be made in the 1930s and subsequently in atomic energy when he postulated that, given an intelligence to handle and sort single molecules, it is possible, without expending any work or violating the principle of the conservation of energy, to warm one half of a mass of gas by cooling the other half. The effect would be to make heat run uphill. Thus modified, the Second Law of thermodynamics becomes one of probability rather than regularity.

Dr Paul R. Heyl of the US Bureau of Standards proved that, in theory at least, here was a perpetual motion machine. He wrote nearly fifty years ago, yet his basic reasoning has now come true. Two physicists, Boltzmann and Planck, working during the closing years of the last century, laid the foundations. Boltzmann showed that the spontaneous equalising of the temperature of two bodies was, in effect, merely the passage of the molecules of these bodies from a less probable to a more probable arrangement. The hypothetical flow of heat uphill was thus seen to be not impossible, but merely improbable.

This can be demonstrated with a simple example. The law of gaseous diffusion is closely similar to the law of flow of heat, diffusion always tending towards a state of uniformity. Differences in density, as with differences in temperature, if left to themselves, tend to disappear rather than to increase. It would be just as surprising to see a gas, originally of uniform density, crowd itself spontaneously into half the volume, leaving the other half as a vacuum, as it would be to see heat run uphill.

Now let us suppose a volume of gas so small that it contains but two molecules, one in each half of the volume. These molecules are continually in motion, rebounding from the containing walls and passing back and forth irregularly from one half of the volume to the other. We may have four possible different arrangements:

$$A \quad B; \quad B \quad A; \quad AB \quad O; \quad O \quad AB$$

Of these four possibilities, two represent a vacuum in one half of the space. The probability of such an occurrence is therefore $\frac{1}{2}$, or in other words it may be expected to occur half the time.

As the number of molecules is increased, the probability of a vacuum arising in this manner decreases rapidly. For a total of n molecules, the probability that half the volume will be empty is $\frac{1}{2}^{n-1}$, which, for the enormous number of molecules normally dealt with, is, to all intents and purposes, *zero*. For the much more probable case that the pressure in the two halves of one cubic centimetre of a gas should differ by one per cent, the probability is still so small that such a condition may be expected to arise spontaneously only about once in $10^{10^{18}}$ years!

These sound pretty long odds, but one point needs clarifying. If an occurrence is so rare that it may be expected to happen on the average once in an aeon, this does not necessarily mean that we will have to wait that long for it to happen. There is nothing to prevent its happening in five minutes' time and the probability of that happening then is exactly the same as the probability that it will happen at the end of an aeon. Again, it might even occur twice in the next minute, although if it did time itself might not be long enough for it to happen again.

Dr Heyl suggested that a similar line of reasoning could lead us to an analogous conclusion with respect to the possibility of the spontaneous appearance of temperature differences. The temperature of a gas we know to be determined by the velocity of its molecules. In a gas at what we consider to be constant temperature, the velocities of the individual molecules are far from being the same, but are distributed about a mean which remains constant. Suppose we now consider a volume of gas so small that it contains only four molecules made up of two equally fast-moving ones, F_1 and F_2, and two equally slow-moving ones, S_1 and S_2. The various possibilities of their arrangement, assuming that there is no change in density, are six in number:

F_1S_1	F_2S_2
F_2S_1	F_1S_2
F_1S_2	F_2S_1
F_2S_2	F_1S_1
S_2S_1	F_1F_2
F_1F_2	S_1S_2

Of these six possibilities, the first four represent cases of equal temperature in the two halves of the gas, for our temperature-calibrating instruments give us only a mean value.[1] The last two, however, represent differences of temperature, the probability of which, for four molecules, is consequently one in three.

If we increase the number of molecules, the probability of any large difference in temperature resulting in this way rapidly diminishes. It must be recognised that in any volume of gas the temperature of which we can control and measure, the temperature of each and every small portion is continually fluctuating above and below the calibratable norm, and overall the gas is no more uniform in temperature than the surface of the ocean is absolutely level.

That detectable differences of temperature may arise in this way is, of course, highly improbable but, and this is the important qualification, it is not impossible. We must, however, recognise the fact that heat not only *can* run uphill, but that it is continually doing so on a very small scale, and that perhaps very, very occasionally we might just observe it.

Karl Christian Planck, the German philosopher (1819–80), stated the case in a succinct way. He said that if a kettle of water be placed on the fire, there is a chance, albeit an exceedingly small one, that the water will freeze!

[1]This assumption holds good for the experimental laboratory.

Once it had been recognised as possible for the uphill transfer of heat and that uphill changes in temperature and density were continually occurring on a very small scale, the question was raised whether or not it was possible by some device to accumulate these changes until a sensible effect could be produced and useful work obtained. Such a device, postulated the physicists seventy and eighty years ago, may be termed a perpetual motion machine of the second kind, since it obtains work not by a creation of energy but in spite of the second law of thermodynamics.

In 1900, Lippmann of Paris proposed just such a device, and in 1907 Svedberg of Uppsala made a similar suggestion. But in 1912 a scientist called Smoluchowski published an extended discussion of the theory of the subject and, as a result, laid down the general principle that we can hardly hope to accumulate these rare departures from the Second Law on a molecular scale by any device built up of molecules, for every such device is itself subject to molecular variations which, in the long run, will probably cancel out the variations that we hope to pick up by its aid. This argument, although by no means final, was certainly discouraging.

Notwithstanding this, it is sufficiently novel and surprising to recognise that the Second Law can be said to be valid only in a statistical sense in the long run.

A strange demonstration that molecular adjustment might occasionally be unequal was reported in 1948. Dr Lewis V. Judson of the National Bureau of Standards presented evidence to the American Physical Society that one of the Bureau's one-metre standards (metal bars used as a standard for dimensions) had systematically been growing longer over a 45-year period. Its length had increased 1·3 microns (52 millionths of an inch) during the preceding six years. A companion bar had been shrinking and was then 80 millionths of an inch shorter than six years previously. Judson told the Society, 'We think it may have something to do with molecular instability. The molecules in the bars appear to be rearranging themselves. But we don't know.' Although companion bars made of platinum iridium alloy had not altered by so much as a split millionth of an inch, the two bars in question, which were both of nickel steel alloy, certainly demonstrated a rare characteristic.[1]

[1] *Science Digest*, Chicago, vol. 24, August 1948, p. 46.

But when Professor Debye addressed the Bureau of Standards back in March of 1925, he warned his audience of scientists that in order to reconcile the then identified phenomena of interference with the quantum concept, it might be necessary to assume that the conservation of energy might itself only be true statistically. Energy, he postulated, might be created and annihilated in minute fluctuations, its average remaining constant in the long term.

Here was the first hint that perpetual motion of the first kind—the actual creation of energy—might be a scientific probability and even a possibility. In a later chapter we will see just how true this was to come.

Although perpetual motion might be described as one of the more recent follies of science, with both alchemy and the quadrature of the circle being of greater antiquity, the century or more during which its quest occupied the minds of men produced teachings far more valuable than the goal which could not be attained. Preston[1] expressed it:

'The alchemists in chemistry have been somewhat like the perpetual motionists in natural philosophy. Both, by seeking after the impossible, have led up to discoveries of the greatest importance and practical value.'

One of the first to conclude that perpetual motion was akin to the experience was Simon Stevinus who was born at Bruges in 1548 and lived until 1620. This great mathematician was also a practical man—among his inventions was the land yacht with which he amused himself and friends on the sea-shores of the Netherlands in about 1600. An early advocate of decimal coinage and decimal fractions for daily use (his system was rather unwieldy), it was Stevinus who distinguished between stable and unstable equilibrium. Of greatest interest to us, though, was his proof of the law of equilibrium on an inclined plane and this he did by showing that perpetual motion could not exist. He supposed a flexible, endless cord of uniform density on which are strung fourteen balls of equal mass and at equal distances apart. This he hung on a triangular-shaped support comprising two unequally inclined planes with a common horizontal base. This is shown in Fig. 2.

Let the portion AEB contain eight balls. For the sake of sim-

[1] *Theory of Heat*, p. 7 (n.d.).

plicity, let AC=2BC. In investigating the condition for equilibrium, one of two things must pertain: either the balls will be in equilibrium when so arranged, or they will not. If they are not in equilibrium, motion will ensue, but this motion cannot change the condition of affairs as there will always be eight balls in the part AEB, four on AC and two on BC; hence once the system starts to move it must continue, that is, it must demonstrate perpetual motion. Stevinus was not only unwilling to accept this, but he deemed it wholly improbable.

He showed this problem, and its solution, in his work on the theory of inclined planes published near the close of the sixteenth century.

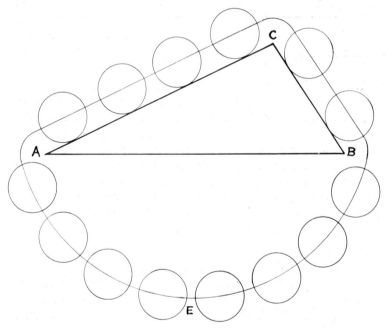

Fig. 2. Stevinus showed that fourteen uniform balls on a uniform chain arranged on a triangle ABC so that four balls lay along AC and two along CB were balanced by the eight balls on the curve AEB.

Stevinus demonstrated how the conditions of equilibrium are not disturbed by removing the eight balls in AEB, since from the arrangement the four balls on the part AE oppose the four on EB. Hence the four balls on the longer inclined plane balance the two on the shorter plane. In other words, the weights would be as the lengths of the planes intercepted by the horizontal plane AB. As shown in Fig. 3, if we take the vertical CB we have but one inclined plane and the condition for equilibrium follows at once.

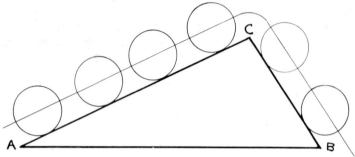

Fig. 3. Stevinus then proved that the balls on AC and CB remained in equilibrium without the need for the arrangement shown in Fig. 4. Were this not so, then perpetual motion would exist.

We have found that the weights are as the lengths, in other words $4 \times 2 = AC \times BC$. Now if we call the two balls the acting force and the four balls the resisting force, then it follows that the resisting force/the acting force = length/height. This is the well-known condition for equilibrium on the inclined plane when the acting force is parallel to the inclined plane.

Ernst Mach, the Austrian physicist and writer on science (1838–1916), was sympathetic to the writings of Stevinus, but perceived that the conclusions were based largely on observations. 'Unquestionably', he wrote in his book *Mechanics*,

'in the assumption from which Stevinus starts, that the endless chain does not move, there is contained primarily only a *purely* instinctive cognition. He feels at once, and we with him, that we have never observed anything like a motion of the kind referred to, that a thing of such a character does not exist. This conviction has so much logical cogency that we accept the conclusions drawn from it respecting the law of equilibrium ... The reasoning of Stevinus impresses us as so highly ingenious because the result at which he arrives apparently contains more than the assumption from which he starts.'

Stevinus, it seems, had succeeded in creating a form of verbal energy!

The next man who appreciated the unreality of perpetual motion was Galileo (1564–1642) in his work of comparing the motion of a body on an inclined plane with that of a body falling freely. His argument is demonstrated in Fig. 4, featuring an inclined plane, AB.

He assumed that the velocity gained in moving from A to B (friction neglected) is the same as the velocity gained in falling freely from A to C. If this is not so, then Galileo showed that the bodies will rise by reason of their weight.

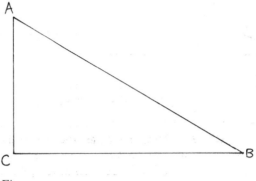

Fig. 4.

Galileo was not altogether satisfied with the mere philosophical discussion of the problem but, being an experimenter of the first rank, he proceeded to test his conclusion by experiment. He fastened one end of a flexible cord to a nail driven into the wall and from the other he suspended a heavy ball. The set-up which he made is shown in Fig. 5.

Lifting the pendulum from M to A in such a way as to keep the tension in the string and then letting go of the weight, he found that it rose to about the same height on the other side. Any diminution he attributed to air resistance.

He then varied the experiment by driving a nail into the wall at point X to the right of the string while it was hanging vertically. The ball now described the arc AM, but when the string struck the nail X, the part CX ceased to move and the ball described the new arc MK. He then inserted another nail below point X at point Y and tried the experiment again. This time the ball described the arc AM as before, and the new arc MG. He observed

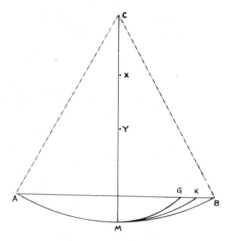

Fig. 5. Huyghens' experimental variable pendulum.

that in each case the ball rose to the same level on the line AB, hence the inclination of the plane was shown to have no effect on the velocity. While the velocity gained in falling from A to C (in Fig. 3) was the same as that gained in moving from A to B, it did not *follow* that the time required from A to B was the same as that required from A to C.

After Galileo, Marin Mersenne (1588–1648) specifically denied the possibility of perpetual motion in the year 1644 and went so far as to place all efforts to construct a perpetual motion machine in the same category as the efforts of the alchemists to find the Philosophers' Stone.

Christian Huyghens (1629–95) was more than likely quite unaware of the work of Galileo when he set out to prove experimentally by the use of *pendula* that the centre of gravity of a free body moving under the influence of the force of gravity cannot rise above its precedent altitude. Huyghens still thought, however, that perpetual motion might be achieved through the use of forces other than that of gravity and seemed particularly to favour magnetism as having some possible bearing on its realisation.

Strange to say, Isaac Newton (1642–1727) because of his failure to make any direct reference to perpetual motion has often been thought to have been ignorant of its impossibility, and in his work on celestial mechanics it is the idea of a central force which is the all-important factor. The concept of energy and work plays but a small role in the *Principia*. In his scholium to the Third Law of motion he lays down the following principle:

The work done by the acting agent is the work done against the resisting agent.

From here to the conservation of energy is but a small step. Concerning Newton's ideas on perpetual motion, Mann in his *Teaching of Physics* makes the following remark:

'It is difficult to conceive that Newton did not perceive that perpetual motion is impossible, especially since Galileo and Huyghens had already made such fruitful use of the intuition. It seems far more probable that he did perceive it, but did not mention the fact, because, with the science of heat still in the intuitive stage, he could not treat it in the same rigorous way in which he treated the other mechanical relations, action and reaction for instance.'

Roget, who was an ardent supporter of the chemical theory of the cell rather than the contact theory of Volta, made intuitive use of the principle of negation of perpetual motion. In his *Galvanism* published in 1832 (subsequently also quoted by Whittaker in his *History of Ether and Electricity*), Roget wrote:

'If there could exist a power having the property ascribed to it by the [contact] hypothesis, namely, that of giving continual impulse to a fluid in one constant direction, without being exhausted by its own action, it would differ essentially from all the other known powers in nature. All the powers and sources of motion, with the operation of which we are acquainted, when producing their peculiar effects, are expended in the same proportion as those effects are produced; and hence arises the impossibility of obtaining by their agency a perpetual effect; or, in other words, a perpetual motion. But the electro-motive force ascribed by Volta to the metals when in contact is a force which as long as a free course is allowed to the electricity it sets in motion, is never expended and continues to be exerted with undiminished power, in the production of a never-ceasing effect. Against the truth of such a supposition the probabilities are all but infinite.'

This discourse was written just a few years before the doctrine of the conservation of energy was formally enunciated.

Carnot used the negation of perpetual motion as one of two principles to establish a relationship between work performed by an engine and the heat drawn from the furnace.[1] His second principle was the conservation of the calorie. Although the caloric principle which he used, described further on, was later to become obsolete in favour of the commonly accepted dynamical theory, Carnot's work remained for a long while an outstanding contribution to our understanding of energy.

About this same time, Faraday used the principle of the negation of perpetual motion in submitting his objections to the contact theory of Volta.[2]

The First Law of thermodynamics, then, says that a fixed amount of mechanical work always produces the equivalent amount of heat, and thus energy can be converted from work into heat, but it can neither be created nor can it be destroyed. This is the dictum in its simplest form and it is this which is the very cornerstone of the principle of the conservation of energy.

The Second Law, stated in equally simple terms, says that heat cannot be increased without the expenditure of more work: it cannot run uphill.

The German physicist Julius Robert Mayer (1814–78) turned his attentions towards establishing the First Law of thermodynamics and, in 1842, formally wrote:

'Once in existence, force cannot be annihilated; it can only change its form.'

Hermann von Helmholtz (1821–94), the German physicist, was able to convince the scientists of the world, at the tender age of twenty-six, that the First Law was a valid assumption when, in 1847, he presented before the Physical Society of Berlin a paper entitled 'On the Conservation of Energy'. He began his analysis by declaring that perpetual motion machines were axiomatically impossible.[3] The validity of a physical axiom can be established instead on repeated observations of nature. Helmholtz did not have to prove his axiom since it was enough to confirm that no one had yet succeeded in building a successful perpetual motion

[1] *The Motive Power of Heat*, 1824.
[2] See *Experimental Researches*, no. 2071.
[3] In physics, as in mathematics, *axioms* are distinct from *theorems*. A theorem is a conclusion that is logically deduced from an axiom. An axiom, on the other hand, does not require logical proof.

machine. Helmholtz set no precedent by this statement since Nicolas Leonard Sadi Carnot (1796–1832), the French physicist and early theoretician on the steam engine, had also begun with a similar axiom, and by this means established numerous conclusions concerning the dynamics of heat. It was Carnot's *Reflections on the Motive Power of Heat* (1824) which formed the basis of the Second Law of thermodynamics.

Helmholtz demonstrated that both heat (considered as small-scale motion) and work (regarded as large-scale motion) were forms of energy and that which was conserved was the total of the two forms rather than either heat or work taken separately.

Stanley W. Angrist[1] rightly commented that some perpetual motion machines do not violate the First Law of thermodynamics in that neither friction nor electrical resistance is a significant problem in their design. They are none the less impossibilities because they attempt instead to circumvent the Second Law of thermodynamics.

The Second Law stemmed from the observations of Carnot and was fully formulated by Clausius. Now the First Law demonstrates that a fixed amount of mechanical work can always be converted into an equivalent amount of heat. This holds true for most machines but not for a heat engine such as a steam engine. The First Law's axiom is no longer true: you cannot completely convert a fixed amount of heat into the same amount of work for the simple reason that some of the initial energy is unavoidably wasted. A steam engine wastes heat (which is energy) through overcoming friction, warming itself and radiating some of that heat into the atmosphere. Other heat is lost through leakage and minor causes.

Carnot made a careful study of the steam engine as a 'closed circuit' in which water was heated until it changed into steam, the steam moved the piston, was exhausted, condensed back into water and fed back into the boiler. He concluded that there was an unavoidable loss of thermal energy in the process of cooling and condensation, and therefore he postulated that the transformation of heat into motive power 'is fixed solely by the temperature of the bodies between which is effected ... the transfer of the caloric'. Carnot's findings stand as the earliest statement of the Second Law, namely that to do work, heat behaves just like

[1]'Perpetual Motion Machines', *Scientific American*, vol. 218, January 1968, pp. 114–22.

water in a water-mill—it must 'run downhill' to do work, and the farther it runs downhill, the greater the amount of work it does. Today we express this by saying that heat must be transferred from a higher temperature to a lower one in order to do work.

Clausius expanded on Carnot's basic work and adopted the term *entropy* to express the scale of heat lost irretrievably. Angrist[1] qualifies:

'The modern formulation of the Second Law that says entropy always increases arises from the earlier realisation that heat is a downhill flow. Because the supply of energy in the universe is a constant that cannot be increased or decreased, and because at the same time the downhill flow of heat is accompanied by inevitable losses, a time will inevitably come when the entire universe will be at the same temperature. With no more hills of heat and therefore, in Carnot's terms, no further transfers of caloric, there can be no work. This inevitable end, sometimes called the "heat death" of the universe, concerns us here because perpetual motion machines that attempt to violate the Second Law are expected to achieve a localised halt in the inevitable increase of entropy and produce a decrease of entropy instead. The fact that, on the average, entropy continually increases does not, of course, rule out the possibility that occasional local decreases of entropy can take place. It is only that the odds against such an event are extraordinarily long.'

This is the same as the case of the kettle on the hob resulting in the freezing of water rather than its boiling. The chemist Henry A. Bent calculated the odds against a local reversal of entropy, specifically considering the probability that one calorie of thermal energy could be converted completely into work. His analogy has gone down in modern literary history for he expressed the probability as akin to that of a group of monkeys hitting typewriter keys at random producing the complete works of Shakespeare. Bent calculated the likelihood of such a calorie conversion as about the same as the probability that the monkeys would produce the Great Bard's works 15 quadrillion times in succession without error!

It was odds such as these which John Gamgee unwittingly faced when he sought to power the US Navy with his heat engine. This is described in Chapter 10.

[1] *Ibid.*

3

Medieval Perpetual Motion

The efforts of the early perpetual motion seekers were seldom committed to paper and although the first printed books dealt with the sciences and the arts, few contributed anything to our quest for early perpetual motion.

A very old Sanscrit manuscript on astronomy called *Siddhânta Ciromani* gives an account of a wheel having on its outer edge a number of holes of equal size and at equal distances from one another, but arranged in a zig-zag—in other words there were two rows of holes staggered around the tread of the wheel. These holes were to be half filled with mercury and then sealed over. It was claimed that if such a wheel were supported on an axis and the axis in its turn supported on a pair of props, the thing would rotate by itself once set in motion. The Siddhânta treatises date from the first half of the fifth century and so are around 1,550 years old.

Later there appeared a strange work called *Antrum Magico-Medicum* written by Mark Antony Zimara. I am indebted to Professor G. Kasten Tallmadge of the Department of Anatomy, Marquette University, Milwaukee, Wisconsin, for his researches into Zimara and his notions and I quote freely from the paper on Zimara which he published in *Isis* in 1941.

Mark Antony Zimara was an Italian philosopher, physician, astrologer and alchemist. He was born at San Pietro in Galatino (Lecce) about the year 1460 and died in Padua in either 1523 or 1532. During his life, he held professorial chairs in both philosophy and medicine in Padua and Naples.

Zimara was a student and critic of Aristotle and of the Aristotelian, Saint Albertus Magnus, whose work on physics and metaphysics Zimara published from Venice in 1518. As a physician he contributed an important work (for its time) on diseases of the body. Learned in the writing of other alchemists, astrologers and thaumaturgists, it is not surprising that large portions of his considerations of disease and of his therapeutics are both spagyric and astrologic.

But what interests us is the fact that the author includes in his treatise a strange perpetual motion machine 'without the use of water or a weight'. The entire work, written in Latin, is devoid of illustration. However, Tallmadge secured the services of a talented modern artist from the Layton School of Art in Milwaukee who prepared a most commendable illustration adhering strictly both to Zimara's description and the styles of illustration contemporary with the original publication. The picture drawn by this artist, Burton Lee Potterveld, is reproduced here as Fig. 6.

Fig. 6. Antony Zimara's self-blowing windmill.

In translation, Zimara's words read as follows:

'DIRECTIONS FOR CONSTRUCTING
a Perpetual Motion Machine without the
Use of Water or a Weight

Construct a raised wheel of four or more sides, like the wheel of a windmill, and opposite to it two or more powerful bellows, so arranged that their wind will turn the wheel swiftly. Connect to the periphery of the wheel, or to its centre (whichever the builder may think better), an instrument which will operate the bellows as the wheel itself turns (this will be an honour to the ingenuity of the maker). It will happen that the wind which comes from the bellows and blows against the vanes of the wheel will cause the wheel to rotate, and that the bellows themselves, operated by the rotating wheel, will blow perpetually. This, perchance, is not absurd, but is the starting point for investigating and discovering that sublime thing, perpetual motion, a starting point which I have not read of anywhere, neither do I know of any one who has worked it out.'

Thus Zimara virtually claims that his idea for achieving perpetual motion is original. The oldest principle probably was that which employed a wheel bearing levers whose lengths were changed automatically by the influence of gravity as the wheel revolved. These are discussed in Chapter 4 below. Of equal antiquity are the machines which made use of the principle of spherical weights rolling upon planes whose inclinations were changed by the rotation of the wheel into which the planes and balls were built. Later, solid weights were replaced by quicksilver (mercury). Other principles which have been used were those dependent on hydrostatic force, capillary attraction and controlled magnetic fields designed to alter the effects of gravity.

However, to Zimara must go the credit of appearing to be the first to suggest the use of the pneumatic principle. That his machine is distinguished for its singular inefficiency is but a detail. The force needed to compress the bellows was far more than that which could possibly be achieved by blowing, with the available amount of wind and its pressure (force), upon a windmill.

Of the accompanying modern illustration, it is worth echoing Tallmadge's words that the landscape, the cloud effects, the curious perspective, the attitudes and dress of the two

'philosophers' and the strange effect of the disproportionate mag-
nification of the ordinary hand bellows, all are authentic to this
kind of representation in the early sixteenth century in Italy. The
'raised wheel of four or more sides' is seen, and 'opposite to it
two or more powerful bellows, so arranged that their wind will
turn the wheel swiftly'. As for the 'instrument which will operate
the bellows as the wheel itself turns', since Zimara left the place
of its attachment to the wheel a matter of choice between the
periphery and the centre, the artist is justified in attaching to the
periphery of the wheel the three levers which constitute this instrument. It can clearly be seen, of course, that the wheel cannot
turn without the levers jamming against each other and so locking
the whole machine. What is wanted is what today we would call a
triple-throw crank. Zimara made no provision for this device, presumably since it could not readily be attached to the wheel, but
would need to be attached to the axis and supported on an extra
bearing. Such a triple crank it is, then, which in Zimara's own
words '*artificis industriæ residebit*'.

Some of the sketches of Leonardo da Vinci (1451–1519) show
cranks and although he published nothing himself, numerous
writers claim to have seen his drawings soon after his death. Geor-
gius Agricola (1494–1555), the physician, published *De Re Metal-
lica Libri XII* in Basle, although the edition did not appear until
a year after his death. Several of the curious illustrations in this
volume show cranks and also flywheels.

What is more important to us is the fact that Agricola illustrates
the double crank of which Zimara appears to have had no knowledge. Agricola also shows an immense bellows operated by a lever
and used for forcing the draught of a furnace. This he describes
as 'a giant edition of the familiar domestic fire-raising instrument'
and is similar to the bellows drawn in the interpretation of the
Zimara specification.

Zimara was a little more explicit about his ideas than many.
Wise men, mystics and eccentrics of the period were often such
complex 'natural philosophers' that their writings combine magic,
witchcraft, alchemy and astrology with their own brand of psy-.
chosis to the point where any attempt at comprehension is
thwarted on every front.

Wolf, in his *A History of Science, Technology and Philosophy
in the 16th and 17th Centuries*, makes an interesting comment link-

ing early thoughts of the flywheel with the enhancement of energy:[1]

'It appears to be suggested that the steadying effect of the flywheel constituted an actual increase of available work. As the Conservation of Energy is a nineteenth-century conception, such a view was by no means improbable or altogether unreasonable, although Agricola made far fewer mistakes in this respect than did most of his literary contemporaries.'

According to the writings of Thomas Tymme, Cornelius Drebble (1572–1634), the Dutch chemist and natural philosopher, devised a machine which represented the revolutions of the sun and moon (could this have been an early form of orrery?) which he exhibited before King James I (who reigned from 1603 until his death in 1625). This device was actuated by 'a fiery spirit' contained within the axis of the wheel.

Edward Somerset, the Marquis of Worcester (1601–67), the man who is thought to have made the first practical steam engine, described in his strange book *Century of Inventions*, published in 1655,[2] a number of contrivances which would appear from his descriptions to be attempts at perpetual motion.

One of these, numbered XXI in the Century, was entitled *A Bucket Fountain*:

'How to raise water constantly, with two buckets only, day and night, without any force other than its own motion, using not so much as any force, wheel, or sucker, nor more pulleys than one on which the cord or chain rolleth with a bucket fastened at each end. This, I confess, I have seen and learned of the great mathematician Claudius his studies, at Rome, he having made a present thereof unto a cardinal; and I desire not to own any other men's inventions, but if I set down any, to nominate likewise the inventor.'

Number LV is entitled *A double Water-Screw*:

'A double water-screw, the innermost to mount the water, and the outermost for it to descend more in number of threads, and consequently in quantity of water, though much shorter than the

---

[1]XXII, 6, p. 505.
[2]An abbreviated edition appeared in 1746. It should be noted that the word 'century' here means 'one hundred'.

innermost screw by which the water ascendeth; a most extraordinary help for the turning of the screw to make the water rise.'

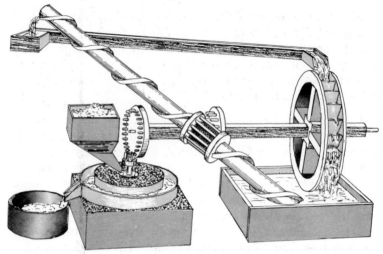

Fig. 7. Closed-cycle mill proposed by Robert Fludd in 1618. Perpetual
power was offered by this means for areas where running streams did
not exist. Not until two centuries after Fludd's death was it understood
that energy conservation made this sort of thing impossible.

There are other references to waterworks which savour of the
influence of perpetual motion, such as how to make a stream ebb
and flow automatically, how to fill and empty tanks of water continually and so on. A most interesting reference concerns a self-
turning wheel and this I quote in the next chapter.

Bishop John Wilkins (1614–72) did not allow perpetual motion
to escape his notice when he was compiling his *Mathematical
Magick*. In this he tells of perpetual motion made by 'chymical
attractions' caused by grinding mercury and some other substances together. By distilling and redistilling the mass, he
assumed that small atoms might be produced which, being placed
in a glass globe, might serve to keep it perpetually on the move.
The worthy Bishop of Chester was, however, wise enough not to
place any great confidence in the efficacy of this proposed process
to secure perpetual motion. I will have more to say on Bishop Wilkins and his inventions in a moment.

The seventeenth and eighteenth centuries were enriched by the
work of engineers, both theoretical and practical, who devoted

their lives to learning and preparing printed works which today give us an insight not just into the engineering capabilities of those far-off days, but into the methods by which work was carried out. Even before the discursive writing of Bishop Wilkins, one great man at least had contributed to our understanding of these burgeoning years. Robert Fludd (1574–1637) left a number of well-illustrated records of contemporary practice and one of these, published in 1618, shows his design for a closed-cycle mill using the overshot waterwheel, the Archimedean screw and a tank of water (Fig. 7).

But it was in Germany that two of the most eminent engineering visionaries lived. The first of these was Georg Andreas Böckler who published a truly remarkable book in Nürnberg in 1686, with a second edition in 1703. This was called *Theatrum Machinarum Novum*, written and illustrated as a record of the progress of the art of engineering. In keeping with contemporary style, the author Latinised his name as Georgium Andream Bocklerum, and described himself as 'Architectum et Ingeniarium'. The work is subtitled 'Exhibens Aquarias, Alatas, Jumentariis, Manuarias, Pedibus, ac Versatiles, Plures et Diversas Molas'.

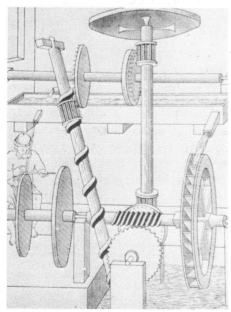

Fig. 8. Almost a century before Fludd, Strade proposed his version of the closed-cycle system, this time as a grindstone with water-cooling, in 1525.

As one might gather, not just from the title page but from a knowledge of the general conditions pertaining in Germany during the century after the Thirty Years' War,[1] Böckler's acquaintance with machinery was restricted almost entirely to mills of one sort or another. In most of these, regardless of the motive power, he depicts the precursor of the geared transmission familiar even to this day.

The seventeenth-century engineers had neither the theoretical knowledge nor the technical equipment to design and shape gearwheels which would mesh with minimum friction. In fact, friction as such, although made use of in such applications as the sacklift in a mill,[2] was little understood. The construction of pinions to mesh with larger gearwheels had yet to assume the form common today. However, the problem of shifting the direction of rotation of a drive through 90° was solved by the invention of the *wallower* driven either by a contrate wheel (a wooden wheel with tooth pegs protruding around the circumference parallel to the axis) or by a cogwheel having teeth projecting radially around the circumference at right-angles to the axis. The wallower was also sometimes called a trundle-wheel and to the clockmaker became the familiar lantern pinion. The wallower comprised two discs of wood, each drilled with a matching set of concentric holes near the perimeter. The discs, usually with square centre bores to aid fixing and to transmit rotary motion, were mounted on a shaft separated by a gap of as much as the mechanism dictated, and threaded through the holes, from one disc to the next, were wooden rods. The result looked not unlike a birdcage or lantern, hence its more common name. When it was meshed with a large wheel having suitably spaced wooden pegs around either its diameter (the contrate) or its periphery (the cog), depending on whether parallel or perpendicular motion was desired, a serviceable gear train was the result. Friction, though, cut efficiency drastically.

All this is depicted in Böckler's *Theatre of New Machines*, but what is a far more striking feature of the illustrations is the constant recurrence of the germ of perpetual motion. The use of

[1]This began in 1618 and after a long and bitter struggle ended in October 1648 with the Treaty of Westphalia.
[2]By tightening a loosely looped rope around a rotating shaft as in a ship's capstan.

water-power seems particularly prone to implant this idea in the human mind. This is probably attributable to the assumption that water comes from nowhere in particular and costs Man nothing. This deludes the miller into assuming that his power costs him nothing by concealing the fact that his power is bought and paid for in terms of units of energy and that it can be delivered to him but once. In any event, it would seem that the proprietor of a water-mill—especially of one whose driving stream was subject to seasonal diminutions of flow—was forever trying to make his water run back uphill and work for him again. Later and wiser mill engineers accumulated their energy when it was plentiful by constructing mill-ponds with sluice gates so that when the natural water flow was diminished, reserves could be drawn upon which did not defy the laws of Nature.

Unfortunately for the peace of the medieval mind, it knew of at least one highly plausible scheme for making water run uphill. If the end of a pipe, coiled like the thread of a screw, is immersed in water, and the whole pipe rotated like a screw, the water will climb up the pipe and keep on climbing so long as the pipe is kept turning. This strange but perfectly workable invention is called an Archimedean screw and is named after the mathematician Archimedes who chanced upon its discovery, and lived from 287 to 212 BC. What we know and understand about the Archimedean screw is that the pipe must be turned by some outside agency. This illuminating piece of information was not understood by our ancestors who, with glinting eyes, asked 'What could be more simple than to connect such an Archimedean screw with the water-wheel of a mill, and make the mill run the screw, and the screw run the mill?' To Böckler, as to many others both before and after his time, the answer was that nothing indeed was more simple. Böckler's mills, which he illustrated in plenitude, all worked on this principle, being provided with a tubful of standing water and he thus laid irrefutable claim to being first and foremost a theoretical genius, and a practical one last if at all. The important thing to keep in mind is that Böckler believed his ideas perfectly sound and one is reminded of the countless subsequent 'geniuses' who through the years have complained bitterly that the stupid dullard of a mechanic entrusted to build their creations has been unable to comprehend their drawings and reproduce them so that the model will do what the sketch so plainly indicates.

One notable feature of the Archimedean perpetual motion machine depicted here is the shape of the motive blades which bear a strong resemblance to the modern turbine (Figs. 9–13).

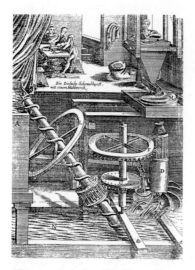

Fig. 9. Georg Andreas Böckler schemed out many self-operating mills using the Archimedean screw. This one uses a coiled tube as a water lift and employs a turbine-like wheel to drive the millstones.

Fig. 10. Another Böckler self-acting mill. This one is even more remarkable since in addition to the screw pump, a crankshaft drives two force-pistons to pump water up the trunk (D) to help turn the drive wheel.

Another means for making a mill raise the water for its own power which Böckler used in several instances consisted of a series of cups attached to an endless rope. The cups were expected to re-deliver the water direct to the wheel.

But it was the self-moving wheel, the subject of a detailed analysis in the next chapter, that appeared the most plausible. Here, with an apparent permanent preponderance of weights on one side of the wheel, once motion was started it seemed obvious that the wheel would not only rotate continuously but would generate enough power to pump water. Böckler's perpetual motion wheel-pump is depicted here in Fig. 10.

The self-moving wheel received special treatment in the writings of the second eminent German philosopher, Jacob Leupold. Leupold published his *Theatrum Machinarum Generale* at Leipzig

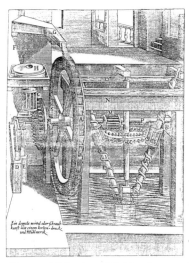

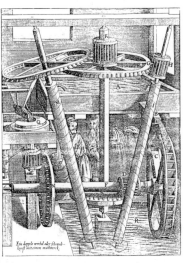

Fig. 12. From water tube to proper helical Archimedean screw, Böckler was obviously dedicated to improving the working life of the miller. But even this one would not have worked.

Fig. 11. This time Böckler has used two water screws plus a crankshaft-driven pump with two square pistons to raise his water. Significantly, some windmill-driven water pumps were built in the Fenland areas of England which had square pistons. However, unlike Böckler's perpetual motion mills, these worked.

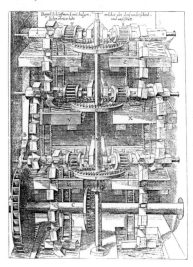

Fig. 13. The most magnificent of all the perpetual motion concepts of Böckler, published in his book *Theatrum Machinarum Novum* in 1703, was this fabulous threefold power unit using triple stages each of double waterlift using a total of six bucket chains. However, one single waterwheel both provided and received the 'power' supposedly available.

in 1724. A number of the mechanisms of Böckler are reproduced here in such a similar style that one must conclude that Leupold drew heavily on the earlier work of Böckler. However, Leupold evaluated the self-turning wheel thoroughly, convincing himself that it must work since the moment of all the weights on one side always (?) exceeded the sum of those on the other side.

We should not be too hard on people like Zimara and Böckler just because we are possessed with the accumulated wisdom and knowledge of many fine brains during the intervening years. We should, on the contrary, commend them for committing to paper a record of the knowledge and technology of their period, at a time when producing a book was a labour of extraordinary dedication. And, just in case we should consider that seventeenth-century technology, or the lack of it, never fooled anybody, Chapter 17 will reveal a remarkable tale about an automatic mill conceived and actually built in Ohio only a century ago.

It is interesting to note that the probability of the Archimedean screw playing any part in the operation of a perpetual motion machine was discarded by none other than Bishop Wilkins in his book *Mathematical Magick* published in 1648 when he was thirty-four years of age and thirty-eight years before Böckler's *Theatrum Machinarum Novum*. The good Bishop of Chester was, in true form for the clergy of his time, both scientist and writer, and when he set about making (and writing about) a perpetual motion machine he went at it as studiously as anyone. The pump was discarded in favour of the Archimedean screw and this he described at great length. He then continued:

'These things, considered together, it will hence appear how a perpetual motion may seem easily contrivable. For, if there were but such a waterwheel made on this instrument, upon which the stream that is carried up may fall in its descent, it would turn the screw round, and by that means convey as much water up as is required to move it; so that the motion must needs be continual since the same weight which in its fall does turn the wheel, is, by the turning of the wheel, carried up again. Or, if the water, falling upon one wheel, would not be forcible enough for this effect, why then there might be two, or three, or more, according as the length and elevation of the instrument will admit; by which means the weight of it may be so multiplied in the fall that it shall be equivalent to twice or thrice that quantity of water which ascends; as may be more plainly discerned by the following diagram [Fig. 14].

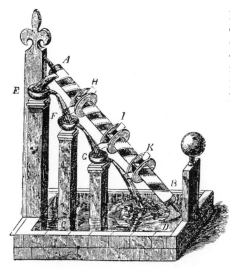

Fig. 14. Bishop Wilkins evaluated the waterwheel-driven Archimedean screw —and finally came to the reluctant conclusion that it just would not work. That was in 1648.

'Where the figure LM at the bottom does represent a wooden cylinder with helical cavities cut in it, which at AB is supposed to be covered over with tin plates, and three waterwheels, upon it, HIK; the lower cistern, which contains the water, being CD. Now, this cylinder being turned round, all the water which from the cistern ascends through it, will fall into the vessel at E, and from that vesel being conveyed upon the waterwheel H, shall consequently give a circular motion to the whole screw. Or, if this alone should be too weak for the turning of it, then the same water which falls from the wheel H, being received into the other vessel F, may from thence again descend on the wheel I, by which means the force of it will be doubled. And if this be yet insufficient, then may the water, which falls on the second wheel T, be received into the other vessel G, and from thence again descend on the third wheel at K; and so for as many other wheels as the instrument is capable of. So that besides the greater distance of these three streams from the centre or axis by which they are made so much heavier; and besides that the ascent of that within is natural—besides all this, there is twice as much water to turn the screw as is carried up by it.

'But, on the other side, if all the water falling upon one wheel would be able to turn it round, then half of it would serve with two wheels, and the rest may be so disposed of in the fall as to serve unto some other useful, delightful ends.

'When I first thought of this invention, I could scarce forbear, with Archimedes, to cry out "Eureka! Eureka!" it seeming so infallible a way for the effecting of a perpetual motion that nothing could be so much as probably objected against it; but, upon trial and experience, I find it altogether insufficient for any such purpose, and that for these two reasons:

1. The water that ascends will not make any considerable stream in the fall.

2. This stream, though multiplied, will not be of force enough to turn about the screw.'

And so Bishop Wilkins not only thought up an attractive perpetual motion scheme, but he also went to the bother of making a model and testing it out sufficiently thoroughly for him to be able to determine that it wouldn't work, and to make a fair assessment as to the apparent reasons why. If only so many other inventors who trod similar paths had bothered to evaluate their ideas in such a manner—and then make their findings public! Robert Fludd (1574–1637) wrote of a scheme similar to this in 1618, but did not evaluate it. He described a waterwheel which sets in motion a chain pump by means of a system of toothed wheels, and the pump is supposed to raise the water necessary to keep the wheel going (Fig. 7).

From this it becomes possible to date the resurgence of perpetual motion in the Middle Ages and to tabulate the authors and their works as follows:

1612 Tymme wrote about Drebble's device which fl. c. 1603–1612
1618 Fludd described his closed-cycle mill
1625 Zimara described his self-blowing windmill
1648 Wilkins evaluated his Archimedean-screw closed-cycle pump
1653 Edward Somerset's inventions

1686   Böckler's self-moving mills
1715   Orffyreus' wheels
1724   Leupold's wheels

From that point on, perpetual motion attempts were legion. What about earlier attempts in this revival? An undoubted classic is that astonishingly imaginative compendium of medieval machinery *Le Diverse et Artificiose Machine* by Agostino Ramelli and published in Paris in 1588. Here we find illustrations of stupendous waterworks for raising water for viaducts, syphons, double-acting pumps, swashplate pumps, low-friction roller bearings, eccentrics, rack-feed drives, automatic reversing gears, Persian wheel water-lifts, a tipping sectional trough for propelling water uphill, and an astounding collection of sophisticated war weaponry including sectional floating bridges, portable hand tools for rending portcullisis and unhinging heavy fortress doors, gunsights, coffer dams and so on. Ramelli was obviously a highly talented engineer, yet nowhere do we find any reference to perpetual motion machinery. Some of the appliances he suggests are close to it, but all are reasonably practical. It is treading safe ground to assert that had Ramelli known of perpetual motion, he would have drawn it and that its absence from his work suggests that perpetual motion in the sixteenth century was at most a rarity.

History has a way of repeating itself and, in a similar manner, inventions have a way of being invented all over again. Bishop Wilkins dismissed the closed-cycle wheel in 1648, yet a variation of this device was put forward with all earnestness by a correspondent to the *English Mechanic* more than 200 years later, the main difference being that mercury was to be used. He wrote:

'In [Fig. 15], A is the screw turning on its two pivots GG; B is a cistern to be filled above the level of the lower aperture of the screw with mercury, which I conceive to be preferable to water on many accounts, and principally because it does not adhere or evaporate like water; C is a reservoir, which, when the screw is turned round, receives the mercury which falls from the top; there is a pipe, which, by the force of gravity, conveys the mercury from the reservoir C on to (what for want of a better term may be called) the float-board E, fixed at right angles to the centre (axis) of the screw, and furnished at its circumference with ridges or floats to intercept the mercury, the moment and weight of which will cause the float-board and screw to revolve, until, by the proper inclination of the floats, the mercury falls into the receiver F, from whence it again falls by its spout into the cistern G, where the constant revolution of the screw takes it up again as before.'

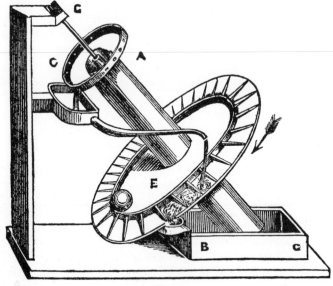

Fig. 15.

Observe how the inventor has chosen a really heavy, labile liquid to impart maximum impetus to his 'float-board'. It is a pity that the same heavy substance must be elevated by the screw in order that perpetual motion should occur, for this we know cannot be done. Had he united the efforts of the alchemist with his thoughts on perpetual motion, then he could have transmuted the mercury into powdered aluminium or, perhaps, a lightweight gas, so as to ascend easily within the Archimedean screw. At the top of the screw, it could then have been reverted quite simply (I am using perpetual motionists' jargon) back into heavy, mobile mercury to spin the screw yet faster in its downwards flow. Sadly such a proposition is only just a shade more impossible than perpetual motion by the means he suggests. For those who chide that one thing may not be more impossible than another, let me remind them of the 'old' saying coined during the last war—'The impossible we do immediately, but miracles take a bit longer...'

Fig. 16. A German perpetual motionist, Ulrich von Cranach, thought up this involved mechanism in 1664. A ball deposited at the top of the paddle-type wheel turns the wheel as it descends until it rolls out and along the semi-circular track at the base. As the wheel turns, it also rotates an Archimedean screw which carries the ball back to the top and sends it down again, so keeping the mechanism in motion. Even the multiplicity of balls and the fine gearing to the screw-lift failed to keep the device in operation.

4

Self-moving Wheels and Overbalancing Weights

It is safe to say that all attempts at perpetual motion in the Middle Ages concerned the wheel which would 'turn of itself'. That the wheel should be so involved in the myth and magic of primitive perpetual motion is not hard to understand. If the stone hammer was Man's first tool, and the tree-branch lever his first appliance, then his first mechanism was the wheel which rotated on an axle. All Man's early machinery centred on the use of the wheel: the horse-driven whin for raising mining spoil or well water, the waterwheel by which corn could be ground and, later, the hammer forge, and the wheels of the other sort of mill driven by the wind. Wheels were at the heart of everything. Lantern pinions, pegged wheels, drives for altering the angle of motion from parallel to anything else, intermittent drives, lifting devices and countless other simple mechanisms depended on the wheel and were derived from its singular uniformity of shape and appearance.

I showed in earlier chapters the way the minds of early engineers and philosophical mechanics worked. One only has to pay a visit to a working windmill today and listen to the creaking and look at the polished pegs on the driving gears to appreciate that creators of wheeled mechanisms had little appreciation of the problem of friction. In fact, many seemed to display an unconscious belief that friction was some sort of a constant to be contended with in any machinery and the way to overcome it was to make everything that much larger so that friction would have less effect. The bigger the machine, and the more wheels, the nearer one came to making a machine which had no friction. Then you only had to add another stage to the system by way of, perhaps, another wheel or water-bucket, and you not only overcame friction completely, but you passed over on to the credit side of the system, and your wheels would turn 'of themselves'.

Far-fetched as this may seem to us today, there was a measure of justification for this belief, and if we can imagine ourselves devoid of the wisdom of the twentieth century, it becomes all too easy to see how perpetual motion could be seen as a natural force, and not so much something to be striven for as something to be used to best advantage.

The German philologist Geiger put forward an interesting suggestion back in the 1870s. He found strong grounds for believing that the Buddhist praying-wheels, on which the prayers of the worshippers were fastened, and which were turned by water-power, were probably the first kind of water motor. There is also an ancient Sanscrit text-book on astronomy called *Siddhânta Ciromani* which seems to contain the first record of a perpetual motion machine. A wheel is described which would 'revolve of itself' (p. 41).

A primitive device for raising water was the so-called *Persian wheel* which, in a flowing stream, would operate automatically and lift water to a higher level. Its utter simplicity and ingenuity must have inspired countless perpetual motionists. There is one point which is the key to its operation: the wheel must be of greater diameter than the height to which the water is to be raised.

The machine comprises a wooden wheel mounted on an axle supported so that it will turn freely. Upon the rim of the wheel are hung a number of buckets on pivots. As the flow of water turns the wheel round, the buckets enter the water and are submerged, and then pass up the rear of the wheel until they come into contact with a fixed trough designed to take a stream of water away from the wheel. The buckets now swing over, discharge their contents and then swing back empty to carry on round with the wheel for immersion again.

A variant of this has no buckets at all, but it has hollow curved spokes. This type was sometimes referred to as De La Faye's pump. It was suitable only for raising water as high as the axis of the wheel, and operated by filling the hollow spokes through holes in the wheel tread. As the wheel revolved, the water was discharged through openings further up the spoke into a trough situated below the axle level. The very old woodcut of a Persian wheel which appears here (Fig. 17) shows both hollow spokes and tipping buckets. Yet another variant was the horn drum, shown in Fig. 18.

Fig. 17. Automatic irrigation was used on the banks of the Nile, the Euphrates, Tigris and all the principal rivers of western Asia using the *Sackiyeh* or Persian wheel.

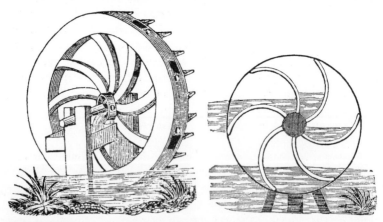

Fig. 18. Two versions of the Persian wheel used either hollow spokes which gathered water through the wheel rim and expelled it through the hub, or comprised a paddlewheel-like assembly of curved spokes to scoop up the water. This was also known as the *horn drum* or *De La Faye's wheel*. Again both operated by the normal current of water in a river.

The Persian wheel was almost completely superseded by 1840 but it served to irrigate land for many hundreds of years without any attention. So long as there was a good flow of water, the wheel turned and did its job. However, it consumed far more energy than its output of work at first sight revealed. A rough assessment of the efficiency of such a machine would, unless it was contained in a mill-race, be very much lower than that of a waterwheel, prob- ably in the region of 15 to 20 per cent. Stagnant water could be lifted in the same manner by driving the Persian wheel by a horse and whin.

This device was a crude version of an instrument known to and described by Vitruvius in 50 BC and even then was derived from Egypt. This was the *tympanum* used for draining land and it was, in a way, a sort of Archimedean screw squashed flat. The French engineer and bridge-builder Perronet used a version of the tym- panum to empty the coffer dams he used for building bridges at Neuilly and Orleans. The same thing, called a *scoop-wheel*, was used in the drainage of the East Anglian fens in the early part of the last century. These, driven by steam engines, proved more effective than the Dutch-style windmills and pumps used earlier in the fenlands.

Yet another version is the *noria* widely used in China and the East in general as late as the last century. These, driven automatic- ally by the water, were frequently very large, and reliable accounts are to be found of wheels 80 and 90 feet in diameter. One, a mere 30 feet in diameter, was built of bamboo, had 20 buckets, lifted 12 gallons per revolution, made four revolutions a minute, and therefore lifted over 300 tons of water a day for the expenditure of no energy other than the flow of the river. Again, the actual efficiency was very low and were it not for the free power of the flowing stream, its cost in terms of energy might have proved pro- hibitive.

All this demonstrates the cult of the wheel and the important part which it played in Man's life and line of thought. So when he evolved the perpetual motion mill, seen in Fig. 13, there could be little simpler and, after all, why should it not work? By repeat- ing his bucket chains, the water could be lifted to a great height, at each level putting yet more power into the central driving shaft, before cascading down on to the waterwheel to keep the whole thing going.

Fig. 19. A variant of the *noria* is this triple wheel system devised by Jacob Leupold. Intended for irrigation, it looks as though it was planned for perpetual motion but in fact is feasible although of low efficiency.

Wheels turned by water and by wind were one thing; the real test of the perpetual motion inventor was to make a wheel which would turn without either.

From the distant date of the Sanskrit astronomical work mentioned earlier, we move forward to the thirteenth century to examine the sketchbook left by the architect Wilars de Honecort. The original is today preserved in Paris. Here we find a drawing of a proposed perpetual motion machine with the statement:

'Many a time have skilful workmen tried to contrive a wheel that shall turn of itself: here is a way to make such by means of an uneven number of mallets or by quicksilver.'

The illustration accompanying this shows four mallets upon what is evidently meant to be the descending side of the wheel, and three on the ascending side, the descending ones therefore overbalancing those on the other. To get the mallets into this desirable position, the top one on the descending side has evidently been made to fall over before its shaft has become vertical. So long as this occurs with every mallet on the upper portion of the ascending side, then the wheel will turn until it wears out. Unfortunately, the only way to do this is to move the mallet by an external force. Once again our chimerical quest has defeated us—and de Honecort.

About two centuries later Leonardo da Vinci produced sketches of six designs, but it is not known for sure whether these were of his own invention or sketched from the work of others. From this moment onwards, the perpetual motionists clamoured with automatic wheels.

Around 1685, the supposedly magical qualities of quicksilver (mercury) were still being applied to the wheel as a means of achieving perpetual motion. The arms or spokes of the overbalancing wheel were hollow and carried a leather bag or bellows at each end with the idea that as the wheel revolved, the mercury would flow from each reservoir as it rose above the horizontal line and pass through to the bellows on the descending side, so accelerating the downwards movement. Perpetual motion, though, remained elusive to the mercury-wheelers and to those who tried the economy version using lead shot. Like de Honecort, they all failed. Or most of them.

One or two strange machines did appear which gave a degree of contemporary credence to their constructors' claims for having found perpetual motion. Before detailing these, though, it should be borne in mind that when these machines appeared, the world was more credible than it is today, and for this reason it would have been less difficult for a clever man to fool many than, say, now.

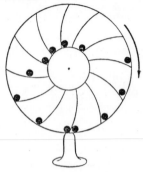

Fig. 20. The oldest fallacy of all is the perpetual motion wheel whose curved spokes throw heavy balls far out on one side, yet draw them in close to the centre on the other. However, the sum of the weights and their moments about the centre is always constant and no motion can result.

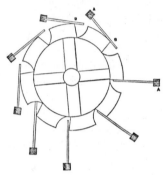

Fig. 21. A variation on the ball and wheel arrangement using hinged mallets.

Fig. 22. Yet another variation on the ball wheel and quite as impractical.

Fig. 23. George Lipton in England schemed up this remarkable device a hundred years ago. The multi-jointed arms each terminated in a cup which carried a steel ball. As the wheel rotated clockwise, the arms closed up and wrapped themselves around the wheel until the ball was conveniently tipped on to a slide which carried it across the diameter to the other side where it was dropped into the cup on the end of another descending arm.

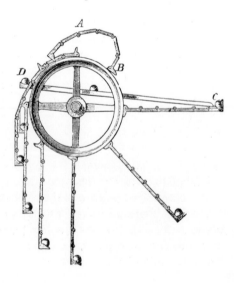

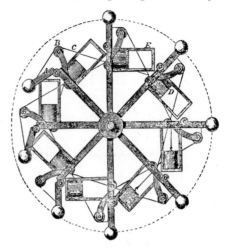

Fig. 24. The English astronomer James Ferguson thought up this scheme where hinged arms carrying weights at their extremities were drawn in and allowed to move out by the force of gravity acting on pistons in cylinders. Note how the movement of one piston is supposed to affect the movement of a weight two arms away via cords and pulleys.

Fig. 25. Weights pivoted on bell-cranks *D*, the short arms of which engage with a fixed guide *G* as the wheel revolves. This turns the lever on its pivot and brings the weighted end on to one of the stops *I* within the periphery of the wheel. A brake strap *K* may be tightened upon the wheel rim by the screw *L* when it is required to stop the wheel.

The first mysterious story of self-moving wheel concerns that described by the Marquis of Worcester in his *Century of Inventions* published in 1655, four years before his death. The item listed as number LVI comprises the following under the title *An Advantageous Change of Centres*:

'To provide and make that all the weights of the descending side of a wheel shall be perpetually further from the centre, than those of the mounting side, and yet equal in number and heft to the one side as the other. A most incredible thing, if not seen, but tried before the late king (of blessed memory) in the Tower, by my directions, two extraordinary ambassadors accompanying his majesty, and the duke of Richmond and duke Hamilton, with most of his court, attending him.

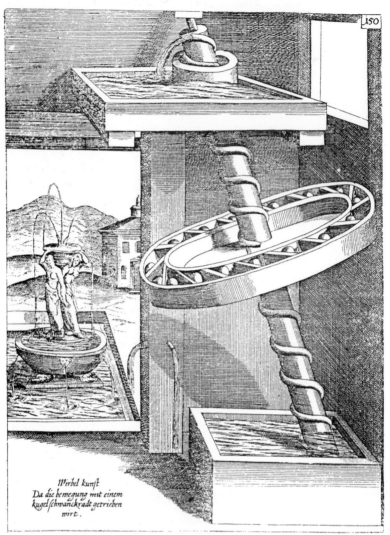

Wirbel kunst
Da die bewegung mit einem
kugel schwanckradt getrieben
wirt.

Fig. 26. Another of Böckler's perpetual motion machines, this time a
water pump to maintain a head of water so that a fountain may play.
The water is lifted by an Archimedean screw which is turned by balls
eccentrically moved in a perpetual wheel.

'The wheel was fourteen foot over, and forty weights of fifty
pounds apiece. Sir William Balfour, then lieutenant of the Tower,
can justify it with several others. They all saw that, no sooner these
great weights passed the diameter-line of the lower side, but they

hung a foot further from the centre, nor no sooner passed the diameter-line of the upper side, but they hung a foot nearer. Be pleased to judge the consequence.'

We have no reason to doubt that the Marquis did construct his wheel and that it was demonstrated as recorded. How, though, the wheel was able to continue turning after the first impulse had been given to it I cannot imagine, always assuming that it did continue to turn. We might judge, though, that a wheel of this size, if accurately made and provided with reasonably low-friction bearings, might well turn for a fair length of time operating as a normal flywheel. A wheel 14 feet in diameter and weighing 2,000 lb. could be expected to appear impressive once set in motion, and would probably turn for long enough to convince a conditioned audience that they were in fact witnessing a self-turning wheel. It might also be mentioned that the flywheel itself was rare at this time: certainly its purpose in engineering was not fully appreciated and its apparent capabilities unknown to the public at large.

Many writers in the past have hinted that the Marquis *invented* perpetual motion, but I have shown that the chimera goes back much further than the seventeenth century. He must have known that the wheel he made fell short of his claim for it, and I am sure that he was able to justify this in his own mind by his reasoning that the wheel ought to work as claimed and the fact that it did not was purely due to the imperfect manner in which it was made. With better workmanship, the wheel would have worked—this is the age-old excuse of the perpetual motionist.

Mechanician to the Marquis was Caspar Kaltoff, and he no doubt put the wheel together in a hurry so as to exhibit it on the occasion of the King's visit to the Tower. Whatever was the true story, the Marquis seems to have abandoned work on his wheel soon after. Years earlier, he had been engaged in perfecting his 'water-commanding engine', forerunner of the steam engine, at Raglan Castle in Monmouthshire. When Charles I was defeated,[1] Edward Somerset (the Marquis) had fled to France and finally returned to England, only to be imprisoned in the Tower of London, during which time it is thought that this wheel was invented. Probably as a result of this, he was freed under Charles II and moved to Vauxhall where he perfected the basic principle of his

[1] At Naseby, 14 June 1645. He was beheaded on 30 January 1649.

water engine under the protection of an Act of Parliament. He
died before practical use could be made of this which, in the final
analysis, was probably his finest achievement. Whereas his self-
moving wheel benefited nobody, the water-commanding engine
had contained the germ of steampower which was to bring fortune
to the nation during the time of the Industrial Revolution.

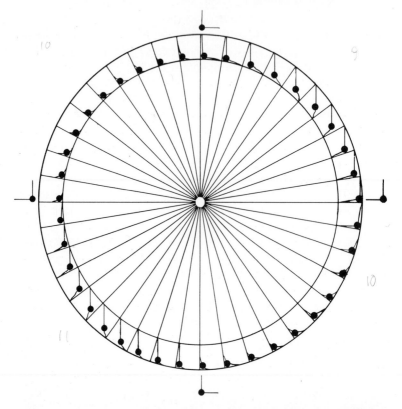

Fig. 27. A reconstruction of the wheel built by the Marquis of Worcester showing how the wheel is supposed to have been maintained in
motion by the swinging in and out of heavy weights, each attached to
a rope.

About the year 1715 a great stir was made on the Continent
by a machine contrived by a German Pole named Johann Ernst
Elias Bessler (1680–1745). A mechanical engineer (they called
them mechanicians in those days), Bessler very soon elected to

drop his surname in favour of the name Orffyreus which he is said to have manufactured by placing his real name between two lines of letters and then picking out alternate letters above and below. Educated for the church, he chose to take up mechanics and soon became an expert clockmaker. His character, according to his contemporaries, was fickle, tricky and irascible.[1] He seems to have developed an early penchant for perpetual motion and experimented with no fewer than 300 different machines until he at last found one, it seems, that worked.[2] However, the government of Hesse Castle imposed a heavy tax upon it and this perhaps understandably led its inventor to break it up.

In spite of this, a second wheel was built[3] and this was shown to the Landgrave of Hesse who allowed it to be erected in a room which was then locked and the lock secured with the Landgrave's own seal. At the end of two months[4] (Hele Shaw quoted forty days), the room was unlocked and the wheel was found still to be turning at apparently the same speed.

Kings, princes, landgraves, not to mention professors and learned men all came and wondered, and went away convinced of the absolute certainty of the action of the machine. This wheel, one of several self-moving wheels he allegedly made, was 12 feet in diameter and 14 inches thick, the material being light pine boards, and the whole was covered with waxed cloth to conceal the mechanism. The axle was 8 inches thick.[5]

One Baron Fischer wrote to Dr Jean Desaguliers, the French natural philosopher, following his examination of it in the castle at Wissenstein in Cassel:

'The wheel turns with astonishing rapidity. Having tied a cord to the axle, to turn an Archimedean screw to raise water, the wheel then made twenty turns a minute. This I noted several times by my watch, and I always found the same regularity. An attempt to stop it suddenly would raise a man from the ground. Having stopped it in this manner it remained stationary (and here is the

---

[1] John Phin, *The Seven Follies of Science.*

[2] Professor Hele Shaw, University College, lecture at Liverpool, 21 December 1887.

[3] 'Perpetual Motion—Some Examples of Misguided Ingenuity', *Scientific American*, vol. 105, 18 November 1911.

[4] *Ibid.*

[5] John Phin, *The Seven Follies of Science.* But see Professor s' Gravesande who stated it to be 6 inches.

greatest proof of a perpetual motion). I commenced the movements very gently to see if it would of itself regain its former
rapidity, which I doubted; but to my great astonishment I
observed that the rapidity of the wheel augmented little by little
until it made two turns, and then it regained its former speed.
This experiment, showing the rapidity of the wheel augmented
from the very slow movement that I gave it to an extraordinary
rapid one, convinces me more than if I had only seen the wheel
moving a whole year, which would not have persuaded me that
it was perpetual motion, because it might have diminished little
by little until it ceased altogether; but to gain speed instead of
losing it, and to increase that speed to a certain degree in spite
of the resistance of the air and the friction of the axles, I do not
see how any one can doubt the truth of this action.'

The wheel was apparently stopped to save undue wear, but the
inventor kept his secret very close. The Landgrave gave Orffyreus
'a fine present' and was then shown the interior, having first
promised not to tell what he had seen nor to make use of his knowledge. In fact, Orffyreus demanded a sum equal to about £25,000
for his secret, a price which apparently nobody was prepared to
pay.

Orffyreus himself wrote several pamphlets extolling the virtues
of his wheel[1] and exhibited the wheel working hard, raising and
lowering stones or water as required. People were not allowed
access to the wheel which, we are told, remained locked in a room,
but they could see the work supposedly done by the wheel by
means of a rope which passed through an opening in the wall.
This, thought the inventor, should satisfy them. Yet there were
disbelievers, among them a M. Crousaz who wrote:

'First Orffyreus is a fool; second it is impossible that a fool can
have discovered what such a number of clever people have
searched for without success; third, I do not believe in impossibilities ... fifth, the servant who ran away from his house for fear
of being strangled, has in her possession, in writing, the terrible
oath that Orffyreus made her swear; sixth, he only had to have
asked in order to have had this girl imprisoned, until he had time
to finish this machine ... eighth, it is true that there is a machine

---

[1]One in Latin, entitled 'Triumphans Perpetuum Mobile Orffyreanum', and one
in German, 'Das Triumphirende Perpetuum Mobile Orffyreanum'.

at his house, to which they give the name of perpetual motion, but that is a small one and cannot be removed.'

These strange comments might be taken to imply that there was something not altogether right with Orffyreus' perpetual motion machine, not to mention the fact that his preparation for the Church appeared to have been sadly ineffective. These suspicions were never confirmed, but they were strengthened somewhat by the events which followed. The Landgrave employed Professor s' Gravesande of Leyden (1688–1742), the Dutch philosopher and engineer, to investigate the mysterious wheel in so far as he could without opening it up. The Professor subsequently wrote to Sir Isaac Newton in which he described it as made of

'several cross pieces of wood framed together, the whole of which is covered over with canvas, to prevent the inside from being seen. Through the centre of this wheel runs an axis of about six inches diameter, terminated at both ends by iron axes of about three-quarters of an inch diameter upon which the machine turns. I have examined these axes, and am firmly persuaded that nothing from without the wheel in the least contributes to its motion. When I turned it but gently, it always stood still as soon as I took away my hand; but when I gave it any tolerable degree of velocity, I was always obliged to stop it again by force; for when I let it go, it acquired in two or three turns its greatest velocity, after which it revolved for twenty-five or twenty-six times in a minute. This motion it preserved some time ago for two months, in an apartment of the castle; the door and windows of which were locked and sealed.'

The Professor certainly appears to have had some measure of faith in the wheel and the demonstration of its ability to turn without apparent external force. We ought not to forget, though, that it may have proved easy to dupe an honest old man whose confidence in humanity was probably unbounded. Orffyreus, however, when he learned that his wheel had been the subject of an inspection, even though only an external one, became so irritated that he once more smashed his wheel into pieces and left a message on the wall saying that he had been forced to do this by the impertinence of the good Professor and, by inference, the Landgrave himself. It is not recorded whether the ageing academician ever received a reply to his letter from Sir Isaac Newton, nor is anything further known of Orffyreus and his strange wheels.

There seems little doubt that the Orffyreus wheel was a fraud being driven by some mechanism in the very large diameter axle. As already determined, its builder was at one time a clockmaker and even as early as this, clocks could be made to run for a lengthy time without rewinding. Forty days, then, was not a long time. Properly balanced, the momentum of a wheel of this size would certainly allow an impressive show of work at intervals, *particularly if the drive was taken from the axle*!

A slightly more complex self-moving wheel was that contrived around 1790 by Dr Conradus Schiviers. This consisted of a wheel with compartments similar to those of a waterwheel, and an endless chain passing over pulleys. A number of balls were placed in a trough at the top of the wheel whereupon they ran, one by one, down an inclined plane into each compartment of the wheel, so causing that side of the wheel to descend. As each ball arrived at the bottom of the wheel, it was discharged into one of the buckets on the endless chain which then lifted the ball back up to the trough at the top. The wheel moved the chain, and the chain would move the wheel. It was a well-engineered scheme, but that didn't overcome its disinclination to turn of itself.

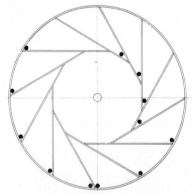

Fig. 28. Leupold's perpetual wheel using balls sliding along straight paths apparently in a frictionless manner. Leupold calculated that the moments of the weights about the centre would always admit of a preponderance on the right-hand side—always assuming that the ball on the horizontal section on the right-hand side could be persuaded to roll from left extreme to right extreme the instant the sector became horizontal.

Of the same vintage was the device put forward by John Haywood. This wheel had a cranked axle to the end of which several rods were connected, passing from side to side across the main wheel, and terminating in small wheels which bore on the inside of the rim of the main wheel. Haywood considered that by this arrangement there would always be a preponderance of weight on the downward side—the side towards which the wheel spun. The points at which the connecting rods were supported by the crank were continually changing so as to constitute levers whose longer arms would always be on that side in every position of the main

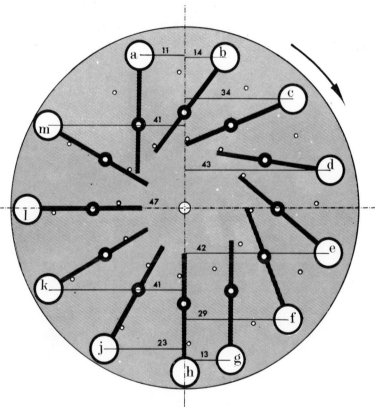

Fig. 29. Leupold demonstrated the same principle using weighted arms. The critical one here is lettered 'a'. It is impossible for this one to adopt the vertical position shown at this point in the revolution of the wheel and unless this can occur, Leupold's calculations are to no avail.

wheel. Here was an instance of proposed sustained motion to be gained from two parts reciprocally maintaining the motion of each other. The perpetual motion remained immobile.

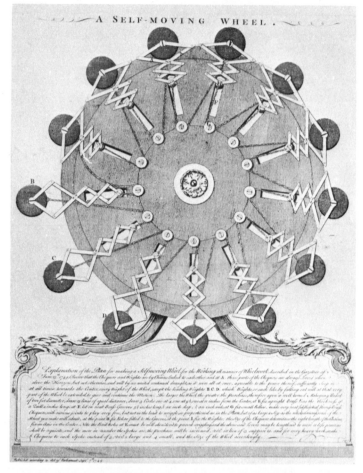

Fig. 30. This complicated scheme for a self-moving wheel was published in September 1749, having been described in the *Gazetteer* for 25 June that year. Its inventor gives complete instructions for making it and ends up with the advice: '... levers may be length'ned to more or less power, as shall be requisite, and the more in number the spokes are, the purchase will be increased, viz in lieu of 13 suppose 21, and for very heavy work, make 6 Chequers [weights] to each Spoke instead of 3, viz 2 large and 4 small, and the size of the wheel accordingly'.

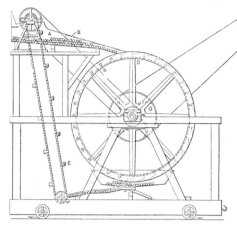

Fig. 31. By the latter decades of the last century, perpetual motion motors such as that shown were being devised by inventors everywhere. This one uses a waterwheel-like wheel fed with heavy balls, an ample supply of which is provided. The power developed by the motor is taken off by belt to the pulley at the top right ...

Fig. 32. Here the balls are dropped into a rotating worm lift after they have performed their function to cause the righthand side of the endless chain to move downwards. Gearing keeps the screwlift turning and drops the balls back into the cups at the top. The governor at the top is to prevent the machine running too fast ...

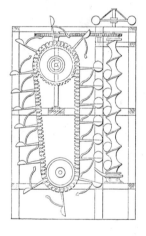

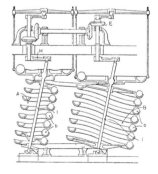

Fig. 33. An ingenious system using two spiral ball tracks, each of different diameter, pivoted at their lower ends about axes *l*. The upper ends of the axes are connected to cranks *H*. Bevel gearing connects the cranks so that the motion stated to be given to the track *A* as heavy balls *b* roll down it, is transmitted to the track *B* to raise the balls again. The balls roll from one track to another along horizontal troughs. The large balls near the top of each track also serve as flywheels.

This is but a small handful of the many early attempts to make self-moving wheels, much the most common among the instruments of perpetual motion, and there were very many of them both before and since. One interesting overbalanced wheel was built as a fraud by a man named E. P. Willis of Connecticut. Angrist[1] considers it perhaps the most elegant wheel ever built. A large toothed gearwheel was mounted at an angle to the horizontal and fitted with a complex system of weights and rods. This purportedly drove a smaller, hollow flywheel. The whole was mounted in a frame and placed in an apparently sealed glass case (Fig. 34).

Fig. 34. Connecticut engineer E. P. Willis made this elegant over-balancing wheel perpetual motion machine in the 1850s. Actually the slender strut *A*, visible centre left, forced compressed air against the wheel which in turn drove the angled wheel with its shifting weights.

The machine attracted a great deal of attention during its exhibition at New Haven where Willis charged admission for

[1] Stanley W. Angrist, 'Perpetual Motion Machines', *Scientific American*, vol. 218, January 1968, pp. 114–22.

viewing it. He then moved it to New York in 1856. There it was examined by a patent attorney who in the year 1871 commented with asperity that inventors submitted one or another variant on Fludd's closed-circuit water-mill to him every year. The exhibitors, he noted, were careful not to claim that Willis had achieved perpetual motion, but what they did do was to challenge any visitor to provide another explanation for the machine's motion. Although the glass case kept visitors from inspecting the machine too closely, the attorney noted that there was an unusual and apparently non-functional strut placed close to and just below the edge of the hollow flywheel. Ultimately the truth emerged. A steady flow of compressed air, undetectable from outside the case, kept the flywheel turning, for it was the flywheel that drove the overbalanced wheel, rather than the other way round.

A variation on the wheel with overbalancing weights was the wheel driven by a pendulum. Some of these were electro-magnetic devices, but all relied on the swing of the pendulum to turn the wheel incrementally with a form of ratchet escapement, with the wheel providing sufficient energy to keep its pendulum swinging to and fro.

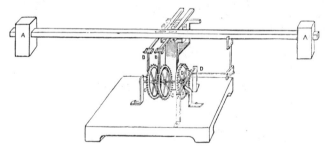

Fig. 35. Leaton's oscillating perpetual motion, 1866.

When a reader of the *English Mechanic*, Mr W. Leaton, chose to write to the editor enclosing details of a machine he invented, he carefully avoided any mention of those two words, perpetual motion. His letter, published in August 1866 and accompanied by the drawing reproduced here as Fig. 35, read:

'Sir,—Permit me to lay before the notice of your readers the accompanying sketch of self-sustaining power, for the opinion of such as may be interested in such a branch of mechanics. I submit that a machine made after this plan will maintain itself in motion, and likewise propel suitably arranged works. The motion is applied by the oscillation of the two balance weights A, A, by alternately pushing down the arms B, B, which communicates a circular motion to the ratchet wheels C, C, the loss sustained by the balance at each movement being overcome by the impetus given it at each movement by a kind of verge escapement movement at D. Should any subscriber detect any error or miscalculation, the inventor of this plan will feel grateful for his opinion thereon.'

The industrious Mr Leaton had, in effect, constructed a see-saw pendulum but, unlike the familiar pendulum of a clock, this one was not being used to regulate power, but to provide it. Because the swinging bar bearing the weights A and A appears to have two pivot points, so as to impart some rotation to the ratchet wheels B and B, rather more weight than is available is momentarily required to swing the high side down. This extra weight must now be removed to the other end for the same purpose there. The unusual escapement serves some strange purpose in the inventor's mind: in truth it can do nothing but add a little more friction and reduce still further the chances of the thing moving. As for exerting 'impetus' at each stroke, where does the impetus come from?

Its originator, to his credit, demonstrated a willingness to be corrected and this certainly was more than many of his fellow perpetual motionists were prepared to invite.

An Australian, reported the *English Mechanic* for 16 November 1900, claimed to have discovered the 'secret' of perpetual motion and was said to have had at his residence a model 'gravity wheel'

'which has concentric rings secured to arms radiating from the centre. Those arms or weights are supposed to act as a balance and driving levers, and are so nicely arranged, that those on the downward grade being two-thirds heavier than those on the upward grade in action, that a natural falling of the weights on the former grade insures perpetual motion.'

After the fiasco of the South-Sea Company formed in 1711 by the celebrated Harley, Earl of Oxford, which went down in the history of monumental failures under the soubriquet *South-Sea Bubble*, an Act was passed by the Lord Justices of Whitehall dated

12 July 1720. This set out to rid the country of spurious investment companies which generally sought solely to deprive the rich and the foolish of their capital. Among the bubble companies was one entitled 'For a wheel for perpetual motion. Capital, one million.' The Lord Justices apparently had a better understanding of perpetual motion, certainly by the wheel, than their fellow men.

Fig. 36. Perpetual motion by oscillation was the aim of this machine. A central pivoted bridge *D* carries a ball *X* which is capable of rolling on the bridge, and is connected by a telescopic tube *G* to the axis of two wheels *M*. These wheels are connected to a pair of balls *N* which run on the curved track *I*. When the ball *X* rolls to the position shown by dotted lines, the balls *N* roll up the track *I* and the

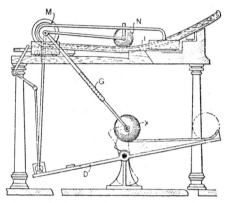

bridge *D* descends on the right. The balls *N* then roll back, bring back the ball *X*, and cause the bridge to swing down on the left. The cycle then starts all over again.

Fig. 37. Charles Batcheller of Iowa patented this mechanism in 1870. The double framework carries two shafts driven together in opposite directions by the meshing gears *G* and *G*. Each shaft carries on it a central fixed gear and a cross arm *B* which is provided with a meshing gearwheel at each end. Suspended from the pivot points of these terminal gears are weights, peardrop-shaped and marked *E*. The cross arm on the

first shaft is at right-angles to the cross arm on the second shaft. The pendulum weights are intended to operate in the usual manner, namely affording more purchase on one half of their revolution than on the other and the resulting power can be taken off on either side of the machine by the gears *C* or the pulleywheel *D*.

Fig. 38. Horace Wickham Jr. of Chicago contrived this version of the oscillating beam machine in 1870. A ball travels along the tube *C* and drops into a spring chute. The beam tilts until it touches a sprung support which opens an internal trap-door and sends the ball off down to the other end of the tube to repeat the procedure. The tilting of the tube drives a crankshaft carrying a flywheel at one end and a centrifugal governor at the other.

Fig. 39. The most graceful and simple perpetual motion machine of all time! An American named F. G. Woodward believed that by mounting a heavy annular wheel between two rollers, one half of the wheel must always be heavier than the other. Sadly, it wasn't.

Fig. 40. Another attempt at making a wheel turn by itself was this one. An endless chain passes over two pulleys *BB* and round two guide wheels *C* and *D*. The upper pulley *C* is thus made to rotate in the opposite direction to the flywheel and the shaft of the pulley rotates so as to drive the edge of the flywheel. The rotation of wheel *C* was supposed to turn the flywheel which in turn turned the chainwheel. This system was not even distinguished by subtlety and was foolish through and through.

Fig. 41. This wheel had hollow spokes with openings in one side at both hub and rim. Steel balls rolled into each spoke at the hub moved outwards as the wheel turned and came out of the spoke at the rim where they were conveyed on a moving belt back up towards the hub to drop into another spoke hole. A Lancashire man, Dixon Vallance Liberton, thought up this way of keeping all his weight on one side of the wheel. Still, though, the weight of one ball had to raise a number of other balls on the belt.

Fig. 42. Doctor Conrad Schwiers thought up his ball-powered wheel in 1790. Once more, the balls entered the paddlewheel at the top, fell out at the bottom and were then lifted straight back up to the top.

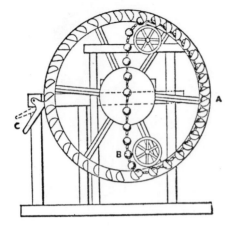

Fig. 43. Here the balls are attached to each other on a chain, but otherwise operate exactly the same as Conrad Schwiers thought of a century earlier. The strange lever *C* is a brake to prevent the engine running too fast.

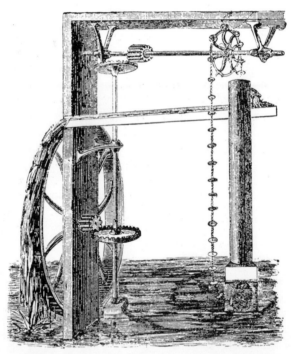

Fig. 44. For comparison, here is one of Robert Fludd's schemes of 1618. The endless rope carries wooden discs which just fit inside the vertical tube at the right. As the chainwheel rotates, one side of the chain moves up the tube carrying with it water as far as the trough which deposits it on the waterwheel. The waterwheel turns the chain through gearing and so motion is perpetual—so Fludd had hoped.

5

Lodestones, Electro-Magnetism and Steam

Perpetual motion by electricity or by magnets stems from a very early date. Magnetic attraction as a characteristic of some materials was recognised at the dawn of Man's experience with ferrous metals. The natural magnet, an ore of iron, consisting of the protoxide and peroxide of iron in a state of combination, was known as a 'lodestone'. The word is said to have been derived from the verb *lead* and *stone*, although old orthography shows that its accepted form was *lodestone*. In ancient times, as now, lodestones were found in considerable quantities in iron mines in Germany, Sweden, Norway, Spain, Italy, China and several other countries. Dark grey in colour and possessed of a metallic lustre, its usual form is a regular octahedron and its specific gravity is 4·25. The lodestone was revered for its special powers in ancient times and its ability to impart a measure of its properties to iron and steel by contact (this was called 'touching') was also known. Articles so touched were known as 'artificial magnets'.

Significantly one of the earliest attempts at securing perpetual motion was an attempt to use this natural phenomenon. It was devised by Peter Peregrinus who lived about the middle of the thirteenth century. The name of Peregrinus is sometimes associated with that of Roger Bacon who was probably the greatest of all the scientists of the Middle Ages. Peregrinus had become familiar with the properties of the lodestone and decided that this could be used to form the basis of a perpetual motion machine. However, he was obviously possessed of rather more wisdom than many of his subsequent fellow perpetual motion seekers, for he stopped short at the idea of the thing and left the task of carrying

out the design to another party. A detailed account of the work
of Peregrinus is to be found in Benjamin's *The Intellectual Rise
in Electricity*.

It was in the year 1269 that Peregrinus wrote his famed epistle
on the magnet in camp at the siege of Lucera. This wonderful
fragment is marred as regards its scientific credibility by a section
of *perpetuum mobile*. Here he proposed a toothed iron wheel to
respond to a magnet in such a way that each tooth was drawn towards the lodestone and then, in some inexplicable way, repelled
so that the wheel turned and the next tooth came up for attention.
Peregrinus placed reliance on sufficient impetus being achieved
to carry the movement over the crucial gap that always renders
the cycle otherwise incomplete.

In 1570 a Jesuit priest, Johannes Taisnierus, thought up a lode-
stone-powered system which is on record as being the earliest perpetual motion scheme of this type for which an illustration exists.
Taisnierus placed his lodestone on top of a pillar and arranged
an inclined plane from ground to pillar top up which iron balls
were to be drawn by the lodestone. I have already mentioned the
work of Bishop Wilkins, *Mathematical Magick* published in 1648.
The Bishop examined the schemes of Taisnierus which he had
published some time prior to 1579 in a book devoted to *Continual
Motions* and to the solving of the problem by magnetism. Wilkins
selected Taisnierus' pillar and supplied the following interpretation:

'But amongst all these kinds of invention, that is most likely,
wherein a lodestone is so disposed that it shall draw unto it on
a reclined plane a bullet of steel, which steel as it ascends near
to the lodestone, may be contrived to fall down through some hole
in the plane, and so return unto the place from whence at first
it began to move; and, being there, the lodestone will again attract
it upwards till coming to this hole, it will fall down again; and
so the motion shall be perpetual, as may be more easily conceivable
by this figure [Fig. 45].'

He went on to explain that while the lodestone might not have
sufficient power to draw the ball straight up from the ground, it
might well do so by the help of the inclined plane.

An examination of the system will show that for the magnetic
force of the lodestone to be strong enough to draw the ball up

Fig. 45. The lodestone draws a steel ball up the sloping ramp until it drops through the hole near the top and rolls back to the bottom where it passes through another opening leading back to the ramp, whereupon the lodestone draws it back up again.

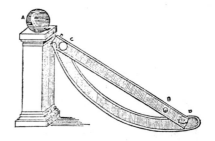

the slope, the ball is hardly likely to be allowed to drop down the hole, let alone return to the bottom of the ramp. After discussing the system in detail, Bishop Wilkins reached the same conclusion that motion of any sort, let alone perpetual, was unlikely. He did, however, add:

'So that none of all these magnetical experiments, which have been as yet discovered, are sufficient for the effecting of a perpetual motion, though these kind of qualities seem most conducible unto it, and perhaps hereafter it may be contrived from them.'

Faith, it seemed, blossomed eternally in the good Bishop's breast.

Magnets sliding in holes in a wooden wheel rotating between two magnets of opposite polarity were proposed by an inventor called Stephan in 1799. Soft iron cores sliding on wheel spokes to respond to the attraction of fixed magnets, so making our old friend the overbalancing wheel into a magnetically attracted over-balancing wheel, cropped up frequently between the early 1800s and the burgeoning years of this century (Figs. 46a and 46b).

Quite a number of inventors devoted their time to trying to find or manufacture a substance which would prevent the passage of magnetic force with the notion that a plate of this substance might be brought between magnet and piece of iron so that magnetic attraction could effectively be turned on and off. A brass wheel carrying sliding iron weights on spokes might then be set in motion between magnets and kept in rotation by bringing such a substance into play at the right moment in the cycle. Various claims to such a discovery were made, the most interesting being that by a shoemaker from Linlithgow named Spence. He maintained that he had found a black substance which intercepted

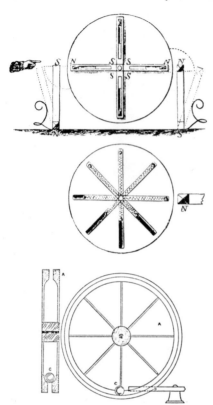

Fig. 46a. Londoner W. Stephan patented this mechanism in 1799. Sliding magnets in slots in the wheel serve as out-of-balance weights to make one side of the wheel heavier than the other. The poles are alternately attracted and repelled by the magnets fixed to hinged supports either side of the wheel.

Fig. 46b. The steel ball *C* is continually attracted to the magnet *B* which is arranged in such a way that a wheel with a slotted rim will rotate clear of it. As the ball rolls treadmill-fashion, the wheel rotates. So believed the inventor who failed to see that gravity and magnetic attraction would cancel each other out.

magnetic attraction and repulsion, and he produced two machines which he claimed moved thanks to the use of this stuff. The fraud was speedily exposed but it is worth remarking that none other than Sir David Brewster (1781–1853), the inventor of the kaleidoscope and a noted physicist, failed to see through the deception and thought the matter worth mentioning in a letter to the *Annales de Chimie* in 1818. In this he wrote 'that Mr. Playfair and Captain Kater have inspected both these machines and are satisfied that they resolve the problem of perpetual motion'. Nevertheless, neither Spence nor his black stuff succeeded in taming the laws of thermodynamics.

An equally impractical quest was undertaken by some to find a means of arresting the effects of gravity. These savants asserted that the successful solution to perpetual motion lay in the gravity interrupter. With the aid of this, the overbalancing wheel could be given a new lease of life. Carried to its ultimate in refinement, all you needed was a plain wheel on an axle, standing half over the gravity interrupter. The unprotected side would then always be heavier thanks to the attraction of gravity, and the wheel could be expected to turn at high speed. Alas! the gravity interrupter remains undiscovered (Fig. 47).

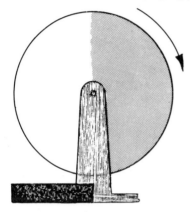

Fig. 47. Simplest perpetual motion scheme of all was to place a piece of gravity-suppressing substance under one half of a freely pivoted wheel. This would instantly make one half of the wheel heavier than the other and continuous motion must result. The problem is to find the anti-gravity material...

As the nineteenth century dawned, science was making its first unsteady steps in electricity—electro-magnetism as it was commonly called.[1]

Science had progressed a long way since the water-mill, and perpetual motionists moved with the times as regards their quest for the elixir of motion. Water had given way to overbalancing weights as seen in the last chapter. Now the magnet and electricity were to come to the fore. In the imperfect state of understanding obtaining at the time of Bishop Wilkins and his iron ball drawn up a ramp by a lodestone, there had been many practical difficulties in trying to use the forces of magnetism. Now, with the skills of clockmaking and machine-tool design, the perpetual motionist could call upon the finest technology to build his mechanisms. Finely balanced wheels in jewelled bearings, finished and polished sliding parts, precision components—all meant less

[1] Although the flow of electricity was discovered as the nineteenth century dawned, static electricity came later, followed by what was properly called 'electro-magnetism'. However, the contemporary term 'electro-magnetism' was used as a general term for each of the many aspects of the burgeoning science.

friction and, so they thought, a surer chance of success. No longer was it necessary to quest for anti-magnets, gravity interrupters or suchlike. The attention of the sophisticated perpetual motionist now centred on the electro-magnet.

One proposition was for a wheel driven by a crank which was connected to the moving contact of an electro-magnet. When the circuit to the magnet was closed, the magnetic attraction was to pull a connecting rod and turn the wheel. As the wheel rotated, two carbon brushes conveyed electricity to energise the magnet. The current was then broken as the wheel completed its revolution, then switched on to make the next revolution and so on. Once turned over by hand, the inventor believed that it would turn until it wore out. Sadly, he had forgotten to make any allowance for the energy lost to friction and to the electrical resistance as well. As with overbalancing wheels, some optimists added springs to help things along, oblivious of the fact that for a spring to pull or push, as the case may be, a great deal more energy has to be expended in stretching or compressing it than will be given out when it is left to its own devices.

The most common electric perpetual motion proposition was the electric motor driving a generator which provided power for the motor. Some were simple, others were complicated, a few were actually projected by talented electro-mechanical engineers. None worked (Fig. 48).

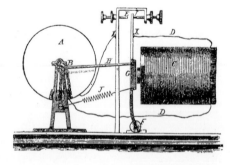

Fig. 48. An American living in Kansas thought up this simple device. Wheel *A* generates electricity and energises the electro-magnet *C*. In the manner of an electric buzzer, core *G* is then moved forward, discharging the coil and in so doing turning wheel *A* by a crank. Speed is regulated by the adjustable contacts. The inventor did not allow for the fact that current would be lost in the leads to the coil and the coil windings so that the power generated would always be less than the power needed to turn the wheel.

It was in the summer of 1902 that the *Daily Mail* announced from one of its correspondents 'A most remarkable claim, the genuineness of which it is as yet impossible to test', made by Señor Clemente Figueras, an engineer of woods and forests (whatever that might be) in the Canary Islands, and for many years professor of physics at St Augustine's College, Las Palmas. I quote:

'Senor Figueras for many years has been working silently at a method of directly utilising atmospheric electricity, and of making practical application of it without the need of employing any motive force. He claims to have invented a generator which can collect the electric fluid, to be able to store it and apply it to infinite purposes, for instance, in connection with shops, railways and manufactures. He will not give the key to his invention, but declares that the only extraordinary point about it is that it has taken so long to discover a simple scientific fact. According to letters received in London from a friend, Mr. E. Ley, of Teneriffe, "Senor Figueras has constructed a rough apparatus by which, in spite of its small size and its defects, he obtains a current of 550 volts, which he utilises in his own house for lighting purposes and for driving a motor of 20 horse-power. Senor Figueras is shortly coming to London, not with models or sketches, but with a working apparatus. His inventions comprise a generator, a motor, and a sort of governor or regulator, and the whole apparatus is so simple that a child could work it." '

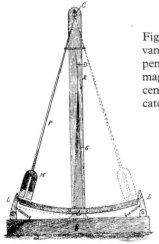

Fig. 49. G. W. Francis of Reading, Pennsylvania, thought up this method of keeping a pendulum swinging forever by the use of a magnet, a rocking curved beam normally centred by a spring at *D* and two spring-loaded catches.

What it was that the esteemed Professor had actually invented, whether it was an electrical storage battery or a strange perpetual motion machine, is, alas, unknown. The 550-volt domestic lighting system certainly must have been something of an achievement, if not just downright dangerous.

For those who need reminding of some of the strange thoughts which prevailed at this time, a French scientist named Taquin, writing in *La Revue Scientifique* in the same year, vociferously attacked those who thought that volcanoes were caused by fire within the earth's crust. He asserted that the only possible hypothesis justified by the whole history of volcanic action was that it was due to electricity! The sun and even the moon were responsible for this generation and, he proclaimed, mountains acted as obstacles to the course of electricity and turned it into fire. The water in the mountains (wherever that may have come from) was then decomposed and rendered highly inflammable. This same man, let it be added, was considered an authority on West Indian volcanoes!

Some interesting and enterprising attempts were made to drive clocks using the natural power within the earth itself. Alexander Bain (1810–77), the Edinburgh clockmaker, discovered that if two electrodes were buried in moist, well-drained soil, a tiny electric current was generated. Bain used the earth battery principle to drive two electric clocks installed in churches in Essex and Suffolk. In the early part of the present century, P. A. Bentley of Leicester produced a number of earth-powered clocks to his own patented design, which featured a system to overcome the difficulties encountered with the Bain type of clock in maintaining constant amplitude of the pendulum which is a prerequisite of good time-keeping.

Bentley produced around 150 of these clocks, one of which ran for 40 years in Leicester Museum without attention. The power came from two electrodes buried three to four feet deep and one foot apart. One was of carbon, and the other was of zinc and between them they provided a potential of approximately one volt. The patent date was 1910 and production ceased in 1914 at the outbreak of war.

These were not perpetual motion instruments so much as extended motion clocks. After a while, the electrodes needed replacement. A great many clocks on both earth- and solar-generated power have appeared from the last century to the present day, but all have a degradable element in their design which removes from them any claim to perpetuity.

Back in the 1870s, a man called Paine exhibited an electro-magnetic engine in Newark, New Jersey. A number of people, including those who should have been educated sufficiently to know better, were inveigled into putting money 'into its development'. Dr Henry Morton went along to view this device one afternoon and reported that the device was shown off to perfection, driving lathes, sawing timber and operating other machinery the moment a connection was made with a small battery of four cells. On the face of it, the machinery was using far more power than could possibly be provided by the tiny electric batteries. Morton spent some while looking at the device but, without taking it to pieces, found himself unable to determine exactly how the fraud, for fraud he was sure it was, was perpetrated. As the afternoon wore on, Paine once again connected up his machine to demonstrate some other aspect. This time, though, instead of bursting into life at full power, nothing happened. The inventor excused the matter by saying that something must have become deranged, and shortly afterwards Dr Morton left. As he did so, he looked at his watch and found that it was five minutes past six o'clock. Within the rest of the rambling building, workshops were filled with machinery driven from a steam engine in another part of the building. And the workers stopped at six. This seemed a strange coincidence, and Morton concluded that the electric battery either served to operate a concealed belt-shift between what was called a fast-and-loose pulley on a steam-powered shafting, or perhaps signalled to a confederate who threw in a clutch somewhere.

Shortly afterwards, some of Paine's investors sought out the inventor and his engine. Both had gone, leaving behind a portion of the iron frame which supported the motor. This had been hollowed out to provide space for a belt drive from the room below. The electro-magnetic engine was—driven by steam!

The steam engine itself was the subject of many claims akin to the family of perpetual motion. One was the aero-steam engine. This was a normal steam engine which included provision for admitting air into the cylinder at part of the stroke. A number of experiments demonstrated, apparently conclusively, that this could increase the engine speed by around 20 per cent. The extension of this was, of course, that you could use air as fuel, or build an air-cycle engine. In truth, it was later found that the boilers supplying these experimental engines were so inadequate that they could not provide sufficient steam for high-speed running, and thus a back-pressure effectively held down the speed. Introducing air at the right point relieved this back pressure and hence the engine ran faster. Subsequently there was a similar flurry of excitement over the projected use of water as fuel. Strange to say, water is used today to augment the power of some aircraft jet engines (it converts to steam in the combustion chambers), but it is not fuel.

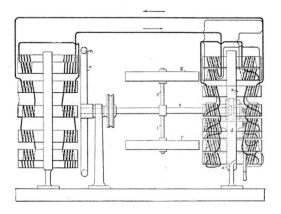

Fig. 50. The magnetic motor was another candidate for perpetual motion. This one was invented in 1894. A series of permanent magnets *a* are mounted on a fixed frame *d* so as to form two semi-circles eccentric to the shaft *e*, and the like poles of the magnets face one way. Two permanent magnets *g* are mounted on arms *h* on the shaft, their similar poles facing towards the magnets *a*. The action of the magnets *a* causes the magnets *g* to rotate as shown in the end view (Fig. 51). The magnetisation of the magnets *a* is increased by means of the windings shown, current being generated in the coils using an armature *n*. This armature is carried by a wheel *m* on the shaft *e* which rotates in front of the magnets *i*.

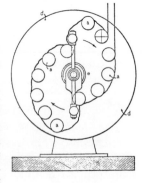

Fig. 51.

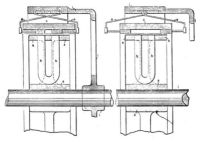

Fig. 52. Another form of magnetic motor is this one where a wheel *a* is fitted with a number of permanent magnets *b*. Around the wheel is a fixed ring *c* fitted with iron blocks sliding in grooves and moved by means of rods *e*, the opposite ends of which run in cam grooves in a wheel *i* mounted loosely on the shaft. On turning the wheel *i*, the blocks slide successively into the positions shown in the two sketches and so cause the wheel *a* to rotate.

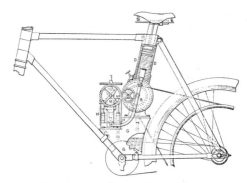

Fig. 53. A boon to the motor-cyclist was how this perpetual motor was described by its inventor. The weight of the rider on the saddle places a pressure on the water in the hollow bar *D*. This forces the water through nozzles and drives a turbine wheel *B*. It then passes to a chamber *G* from which it is pumped back to the bar *D* by pumps *H*. The pumps are driven from the wheel *B* through the spur gearing shown, the vertical spindle *M* having a cam groove into which blocks on each pump rod engage. The cycle is driven by a belt powered by the turbine. Presumably to stop the machine all one had to do was to stand up in the saddle.

6

Capillary Attraction and Spongewheels

As anyone who has left a towel dangling over the edge of a bath filled with water will have learned to his cost, water can flow uphill by what is called capillary attraction. A number of inventors have seen this as a certain means of creating perpetual motion machines. But before relating some of these attempts, it is worth including in this chapter a few of the strange hydrostatic perpetual motion schemes that have been put forward.

The hydrostatic paradox, whereby a very small quantity of liquid apparently balances a very large quantity of liquid, has often been suggested as a means of perpetual motion. That exposed by Denis Papin, the physicist (1647–1712), in the *Philosophical Transactions* for 1685 was virtually the same as that depicted here

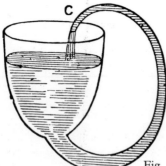

Fig. 54. Boyle's perpetual motion scheme.

(Fig. 54). The goblet, which can be of any substance but which looks better if made of glass, has a hollow stem which continues in tube form around in a curve, rather like a handle, diminishing in cross-section all the while, until finally its narrow open end terminates over the top of the goblet. The somewhat naïve concept is that the larger area of liquid in the goblet must push down and force the underlying liquid into the hollow tube since it weighs so much more. It will finally overbalance the smaller quantity in the narrower part of the tube so that the liquid will spill out of the open narrow end back into the body of the goblet. Once started flowing, the cycle cannot (the inventors firmly believed) be stopped and the liquid will continue to flow round and round until it evaporates. A pint of water in the goblet, they argued, must weigh more than the ounce in the tube. Naturally they were confounded when experiment proved that the water level in the goblet and that in its narrow bent stem were equal.

A similar scheme intended to operate on precisely the same system was proposed by the Abbé de la Roque in *Le Journal des Sçavans* published in Paris in 1686. This instrument was a U-tube with one leg longer than the other and bent over so that any liquid might drop into the top end of the short leg. Here ended the simplicity, for the short leg was to be made of wax and the longer one of iron. Presuming the liquid to be 'more condensed' in the metal than in the wax tube, it was expected to flow from the end of the iron tube into the wax tube and continue *ad infinitum*.

It fell to none other than the famed scientist, mathematician and philosopher John Bernoulli (1667–1748) to demonstrate that whilst most perpetual motion seekers were blighted by the possession of too little learning, here was a case where the reverse was true. Literally translated from the Latin, Bernoulli wrote:

'In the first place we must premise the following [see Fig. 55].

1. If there be two fluids of different densities whose densities are in the ratio of G to L, the height of the equiponderating cylinders on equal bases will be in the inverse ratio of L to G.
2. Accordingly, if the height AC of one fluid, contained in the vase AD, be in this ratio to the height EF of the other liquid, which is in a tube open at both ends, the liquids so placed will remain at rest.
3. Wherefore, if AC be to EF in a greater ratio than L to G, the liquid in the tube will ascend; or if the tube be not sufficiently long the liquid will overflow at the orifice E (this follows from hydrostatic principles).

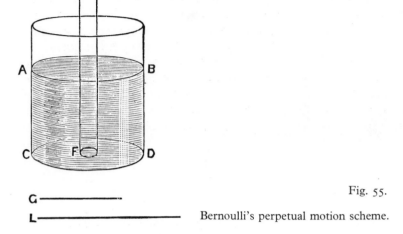

Fig. 55.

Bernoulli's perpetual motion scheme.

4. It is possible to have two liquids of different density that will not mix.

5. It is possible to have a filter, colander, or other separator, by means of which the lighter liquid mixed with the heavier may be separated again therefrom.

'*Construction* These things being presupposed, I thus construct a perpetual motion. Let there be taken in any (if you please, in equal) quantities two liquids of different densities mixed together (which may be had by Hyp. 4), and let the ratio of densities be first determined, and be the heavier to the lighter as G to L, then with the mixture let the vase AD be filled up to A. This done, let the tube EF, open at both ends, be taken of such length that $AC:EF > 2L:G + L$; let the lower orifice F of this tube be stopped, or rather covered with the filter or other material separating the lighter liquid from the heavier (which may also be had by Hyp. 5); now let the tube thus prepared be immersed to the bottom of the vessel CD; I say that the liquid will continually ascend through the orifice F of the tube and overflow by the orifice E upon the liquid below.

'*Demonstration* Because the orifice F of the tube is covered by the filter (by construction) which separates the lighter liquid from the heavier, it follows that, if the tube be immersed to the bottom of the vessel, the lighter liquid alone which is mixed with the heavier ought to rise through the filter into the tube, and that, too, higher than the surface of the surrounding liquid (by Hyp. 2), so that $AC:EF = 2L:G+L$; but since (by construction) $AC:EF > 2L:G+L$ it necessarily follows (by Hyp. 3) that the lighter liquid will flow by the orifice E into the vessel below, and there will meet the heavier and be again mixed with it; and it will then penetrate the filter, again ascend the tube, and be a second time driven through the upper orifice. Thus, therefore, will the flow be continued for ever. Q.E.D.'

Bernoulli then goes on to apply this theory to explain the perpetual rise of water to the mountains and its flow in rivers to the sea 'which others had falsely attributed to capillary action', his idea being that it was all due to the different densities of salt and fresh water.

This wonderful theory, set out so authoritatively and with such adroitness, leaves the reader unable to decide between admiration for such a conscientious statement of the hypothesis, the prim logic of the demonstration, so carefully cut according to the pattern of the ancients, and the massive assumptions built on so frail a foundation.

If Bernoulli believed in perpetual motion by dissimilar liquids, then Robert Boyle (1627–91) entertained the belief that it was possible by capillary action. Boyle apparently recognised in everyday Nature instances where this 'force' was employed as the only way to explain certain natural events. Boyle's reasoning was published in an article in *The Atlas*, subsequently re-published in *The Mechanic's Magazine* in 1827:

'There are many situations in which there is every reason to believe that the sources of springs on the tops and sides of mountains depend on the accumulation of water created at certain elevations by the operation of capillary attraction, acting in large masses of porous material, or through laminated substances.

'These masses being saturated, in process of time become the sources of springs and the heads of rivers: and thus by an endless round of ascending and descending waters, form, on the great scale of nature, an incessant cause of perpetual motion, in the purest acceptance of the term, and precisely on the principle that was contemplated by Boyle. It is possible, however, that any imitation of this process on the limited scale practicable by human art would

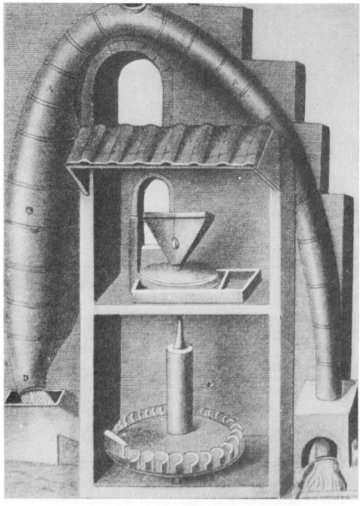

Fig. 56. A variation on Boyle's system was this perpetual motion mill invented by Zonca in 1656. The water is drawn up from the right.

not be of sufficient magnitude to be effective. Nature, by the immensity of her operations, is able to allow for slowness of process which would baffle the attempts of man in any direct and simple imitation of her works.'

Capillary attraction lead Sir William Congreve (1772–1828) to design a perpetual motion system. Sir William, politician and inventor of the Congreve rocket, is supposed to have thought up the idea while convalescing from an illness in 1827, and he considered that it was clever enough in concept to defy refutation on the normal grounds where power is supposed to be derived from gravity alone.

His idea went back to Stevinus and his inclined plane, but there was a difference. Three horizontal rollers were to be mounted in a frame so as to form a vertical triangle with the base horizontal. Around these three rollers there passed a band to which sponges were attached, and around the outside of the sponges was a chain of evenly distributed weights. The whole assembly was then to be placed so that the lower side was immersed in water, so that the waterline reached the centres of the two bottom rollers—the arrangement is illustrated in Fig. 57. The machine was expected to turn in an anti-clockwise direction by virtue of the capillary action of the sponges as follows. On the vertical side, the sponges would be uncompressed by the chain of weights and therefore free to absorb water by capillary attraction. However, on the hypothenuse, the sponges would be compressed by the weights and would

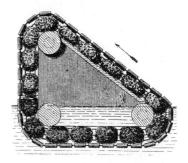

Fig. 57. Sir William Congreve's spongewheel perpetual motion.

therefore have the water squeezed from them. Because the vertical side would therefore be able to soak up more water, motion would ensue.

Sir William went on to calculate the amount of 'power' available with such a machine. He calculated that a fine sponge would soak up water about an inch above the water level. Therefore, if the sponge band were to be one foot thick and six feet broad, the area

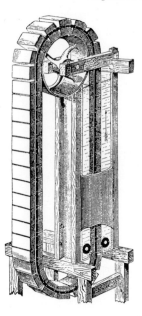

Fig. 58. Philadelphian William H. Chaper devised this spongewheel system around 1870. The tank of water surrounds one half of the endless chain of sponges and assumes the construction of a perfect yet frictionless seal at the bottom where the sponges enter the tank.

Fig. 59. The buoyant motor appeared in many forms, this being the most common layout in which a wheel is mounted so that one half of it is outside the liquid or in a vacuum chamber. The upward pressure on the submerged half produces rotation. The problem of effecting a watertight, frictionless seal is discounted.

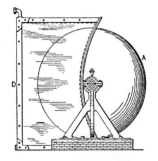

Fig. 60. This buoyant motor features an endless belt attached to which are regularly spaced blades which pick up floats from the bottom of an air-tight tube, rise with them to the top, and replace them at the top of the tube so they drop down and through an air/water valve at the bottom. Air loss and water seepage are counteracted by an air pump which forces air into the tube.

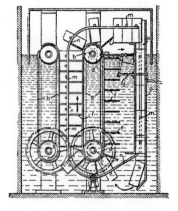

of its horizontal section in contact with the water would be 864 sq. in., and the weight of the accumulation of water raised by the capillary attraction, being one inch rise upon 864 sq. in., would be 30 lb. This, he thought, would more than exceed the friction of the rollers.

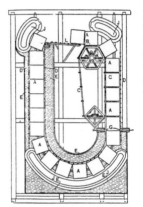

Fig. 61. Similar to the buoyancy motor in Fig. 60, this one drops floats in air down the tube at the right, allows them to pass against the pressure of the head of water (indicated by the water level at the left) through a valve and then do work on turning the motor as they float upwards on the left.

Fig. 62. William Davis of Detroit thought up this version of the buoyancy motor using rubber bags with a weight on the end. As the arms descended into the water the weights kept the bags compressed but once the arms began the upward part of the cycle the weights followed the laws of gravity and distended the bags. There being a hollow air passage right through the arms, this drew the air from the opposite bag into that under water.

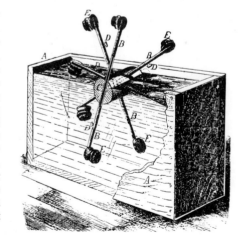

Fig. 63. More rubber bags and balls, this time relying on the displacement of air in a flexible, hollow belt and discounting the friction created by bending such a belt over the two pulleys.

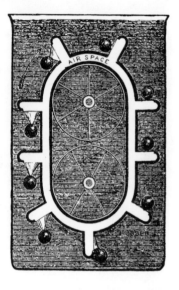

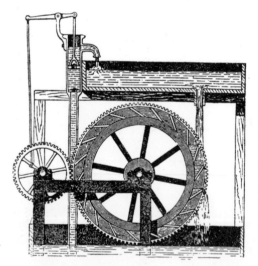

Fig. 64. While the inventors of the late nineteenth century were experimenting forcing air into water, this inventor went right back to the seventeenth century to attempt anew perpetual motion using waterwheel, pump and crank. His success was as short-lived as the duration of his header-tank's contents.

Although the machine was actually patented, Sir William was never able to silence his critics, for his machine failed to stir from equilibrium.

Immersed spongewheels, pneumatic mechanisms where thin rubber bags were inflated with air under water and drawn down again empty on a conveyor belt system, and variations of air pressure and vacuum were all tried. *The Mechanic's Magazine* for 1825 contained a description of an ambitious but highly impractical variation:

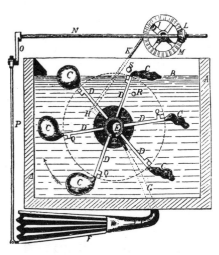

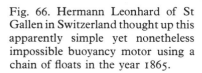

Fig. 65. Rubber bags, belts and cranks again, this time with bellows to force air into each bag as it passed the lowest point in its cycle. A travelling inlet port *E* admits air which is later vented by a valve opened by the fixed peg *S*. The bellows are crank-driven from the rod *N*.

Fig. 66. Hermann Leonhard of St Gallen in Switzerland thought up this apparently simple yet nonetheless impossible buoyancy motor using a chain of floats in the year 1865.

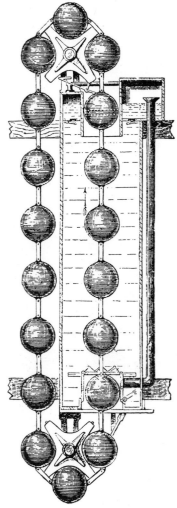

'I beg leave to offer the prefixed device. The point at which, like all the rest, it fails, I confess I did not (as I do now) plainly perceive at once, although it is certainly very obvious. The original idea was this—to enable a body which would float in a heavy medium and sink in a lighter one, to pass successively through the one to the other, the continuation of which would be the end in view. To say that valves cannot be made to act as proposed will not be to show the *rationale* (if I may so say) upon which the idea is fallacious.

'The figure is supposed to be tubular, and made of glass, for the purpose of seeing the action of the balls inside, which float or fall as they travel from air through water and from water through air. The foot is supposed to be placed in water, but it would answer the same purpose if the bottom were closed.

Fig. 67. What could be simpler than a cylinder of water containing floats connected to external weights! This variation on the overbalancing wheel didn't work either.

'Description of the Engraving [shown here as Fig. 68]. No. 1, the left leg, filled with water from B to A. 2 and 3, valves, having in their centres very small projecting valves; they all open upwards. 4, the right leg, containing air from A to F. 5 and 6, valves, having very small ones in their centres; they all open downwards. The whole apparatus is supposed to be air and water-tight.

The round figures represent hollow balls, which will sink one-fourth of their bulk in water (of course will fall in air): the weight therefore of three balls resting upon one ball in water, as at E, will just bring its top even with the water's edge; the weight of four balls will sink it under the surface until the ball immediately over it is one-fourth its bulk in water, when the under ball will escape round the corner at C, and begin to ascend.

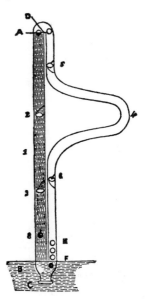

Fig. 68. An 1825 concept of perpetual motion by the interaction of air, water, gravity, valves and balls.

'The machine is supposed [in the figure] to be in action, and No. 8 (one of the balls) to have just escaped round the corner at C, and to be, by its buoyancy, rising up to valve No. 3, striking first the small projecting valve in the centre, which when opened, the large one will be raised by the buoyancy of the ball; because the moment the small valve in the centre is opened (although only the size of a pin's head), No. 2 valve will have taken upon itself to sustain the whole column of water from A to B. The said ball (No. 8) having passed through the valve No. 3, will, by appropriate weights and springs, close; the ball will proceed upwards to the next valve (No. 2), and perform the same operation there. Having arrived at A, it will float upon the surface three-fourths of its bulk out of the water.

'Upon another ball in due course arriving under it, it will be lifted quite out of the water, and will fall over the point D, pass into the right leg (containing air), and fall to valve No. 5, strike and open the small valve in its centre, then open the large one, and pass through; this valve will then, by appropriate weights or springs, close; the ball will then roll on through the bent tube (which is made in that form to gain time as well as to exhibit motion) to the next valve (No. 6), where it will perform the same operation, and then, falling upon the four balls at E, force the bottom one round the corner at C. This ball will proceed as did No. 8, and the rest in the same manner successively.'

This commendably complex piece of reasoning leaves one in no doubt that its inventor was neither a practical man, nor possessed of the faintest appreciation of the characteristics of the elements he was employing—air and water. How he expected a ball

Fig. 69. The French physicist named Ozanam devised this mechanism using in place of air mercury to fill the bellows. He concluded: 'It might be difficult to point out the deficiency of this reasoning; but those acquainted with the true principles of mechanics will not hesitate to bet a hundred to one, that the machine, when constructed, will not answer the intended purpose.'

of 'one-fourth' the weight of water to pass down his air-filled bent tube, opening and closing valves as it went, and finally to open a valve against the air pressure created by his column of water and continue down to the bottom, one cannot imagine.

John Phin quoted a simpler but no less impossible example which he found in the pages of the journal *Power* in the early part of this century. Here a correspondent described a J-tube which was open at both ends but with the lower end tapered or closed in slightly. A well-greased cotton rope passed through the tube and over a pulley mounted some distance above the tube. The curved lower portion of the J-tube (Fig. 71) served as a bottom pulley or guide to the system, and the tapered end closed on to the rope so that it rendered a leak-proof seal between tube and rope. The tube was then filled to the brim with water. The supposed method of working was that the immersed portion of the well-greased rope would tend to rise, while the rope on the other side of the pulley would descend by a combination of gravity and the rise of the rope through the tube.

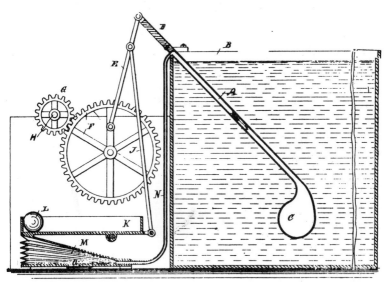

Fig. 70. John Sutcliff of Huntsville, Missouri, actually patented his buoyancy motor in 1882. Crank-operated bellows, aided in their compressive stroke by the weight of a heavy ball *L*, forced air into a rubber bulb when it was under water.

Fig. 71.

The system breaks down once more on the grounds that the mechanism demands the relationship of air and water in an unrealistic and impractical situation. Aside from the friction in the bearing of the pulleywheel, the well-greased rope has to be drawn through a watertight constriction which has to exert some appreciable pressure on the rope in order to retain the liquid in the tube. If one tries to add more 'power' to the lifting side of the system by lengthening the longer side of the J-tube, the weight of water is also increased, the tendency to leak between rope and constriction is increased and so the constriction must hug the rope yet tighter, increasing the friction still further. As if all this was not enough, the friction incurred in bending and unbending the greased rope in the water would also be appreciable. The efficiency of this suggested machine would be considerably less than the value 1·0—and remember that perpetual motion would only ensue if the efficiency were to exceed 1·0. I have already shown that 100 per cent energy cannot, even in an ideal machine, produce much over 85 per cent of work, let alone even 105 per cent!

Combinations of many different liquids have been suggested as being capable of providing just that little bit more efficiency to sponge wheels. I have already described a device using two different media—air and water. Some inventors thought up the most complex of capillary attraction systems which were physically impossible to make. One in particular comprised a continuous belt of sponge passing over two rollers, one placed beneath the surface of liquids in a tank, and the other in free air above. The tank was divided vertically: one side contained fresh water and the other brine and the sponge belt somehow passed through a frictionless but liquid-tight gland betwixt the two fluids. The optimism of the inventor was not daunted by the prospect of fabricating such a device, and he even suggested that the machine might be found to operate *faster* if the tank contained water in one side and liquid petroleum in the other.

Unlike the self-moving wheel experimenters, who usually learned the hard way that their contrivances lacked that certain something, the majority of the capillary attraction brigade appear to have lacked even fundamental experience in physics and engineering. The one exception was probably the spongewheel of Congreve. That, at least, was a jot more practical than the others, and although a failure is a failure whether it be by a fraction or a great gulf, the way an inventor goes about his task must earn him merit marks, even when he is totally wrong!

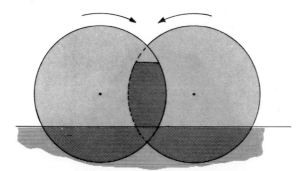

Fig. 72. Capillary attraction between two closely spaced discs would cause one half of each wheel to become heavier than the other. Motion, asserted the inventor of this principle in the 1890s, must ensue.

7

Cox's Perpetual Motion

This book, to a sensitive and maybe sentimental reader, may appear as a chronicle of frustration, of the education of the misguided by painful experience, of wise men driven to despair and of foolish men turned trickster.

But there was one man who made a perpetual motion machine that worked. His name was James Cox, he worked in London, and one can still to this day go and examine his machine.

Cox was a noted clockmaker who made many exotic automaton clocks which were prized, as they still are, around the world. For the story of Cox, of his remarkable museum and some of the marvels which it contained, may I refer you to my book *Clockwork Music*.[1] What concerns us here is what he called his *perpetual motion* which was, in fact, a superbly engineered clock provided with power derived from changes in atmospheric pressure. This he built with the assistance of his chief mechanician, Joseph Merlin, sometime during the 1760s. It was, we are told, the outcome of many experiments to perfect such an instrument.

The main difference between this perpetual motion mechanism and the great majority of those which had passed before (and were to pass in subsequent years) was that Cox did not set out to create energy. This, perhaps, he had the foresight to recognise was an impossibility. What he did do, though, was to make use of a form of energy which was readily available from Nature without human mediation. Cox's concept was a machine which would literally be wound up by natural forces, not by an attendant.

It is necessary to understand that at the time when Cox made his perpetual motion clock and exhibited it in his museum in

[1] Arthur W. J. G. Ord-Hume, *Clockwork Music*, Allen & Unwin, London, 1973.

Spring Garden, it was accepted as normal practice to accompany such exhibitions of ingenuity with descriptive literature of a somewhat flamboyant and mystical nature. This was still the age of mystique and wonder. I emphasise this because Cox's own writings (presented in the third person) might easily be mistaken for the similar announcements put abroad by the charlatans. Cox was, without doubt, a genuine and highly talented man and every piece of his work which survives bears witness to this. One might almost bend the celebrated epitaph of Sir Christopher Wren and say that if you seek a monument to Cox, then look at pieces by him in the finest museums and collections in both hemispheres.

Cox's machine is described in the catalogue of his museum published in 1774.[1] After a lengthy, six-page panegyric, he comes to 'Piece the Forty-Seventh—The Perpetual Motion', and announces it as follows:

'It is a mechanical and philosophical time-piece, which after great labour, numberless trials, unwearied attention, and immense expense, is at length brought to perfection; from this piece, by an union of the mechanic and philosophic principles, a motion is obtained that will continue for ever; and although the metals of steel and brass, of which it is constructed, must in time decay (a fate to which even *the great globe itself, yea all that it inherit*, are exposed), still the primary cause of its motion being constant, and the friction upon every part extremely insignificant, it will continue its action for a longer duration than any mechanical performance has ever been known to do.

'This extraordinary piece is something about the height, size, and dimensions of a common eight-day pendulum clock; the case is of mahogany, in the architectural style, with columns and pilasters, cornices and mouldings, of brass, finely wrought, richly gilt, and improv'd with the most elegantly adapted ornaments. It is glazed on every side, whereby its construction, the mode of its performance, and the masterly execution of the workmanship, may be discovered by the intelligent spectator. The time-piece is affixed to the part, from whence the power is deriv'd; it goes upon

[1]'A Descriptive Inventory of the several exquisite and magnificent pieces of Mechanism and Jewellery, compriz'd in the schedule annexed to an Act of Parliament, made in the 13th year of the Reign of His present Majesty George the Third; for enabling Mr. James Cox, of the city of London, Jeweller, to dispose of his Museum by way of Lottery.' 1774.

diamonds, or (to speak more technically) is *jewelled in every part*, where its friction could be lessened; nor will it require any other assistance than the common regulation, necessary for any other time-keeper, to make it perform with the utmost exactness. Besides the hour and minute, there is a second hand, always in motion; and to prevent the least idea of deception, as well as to keep out the dust, the whole is enclosed within frames of glass, and will be placed in the center of the Museum, for the inspection of every curious observer.

'N.B.—The very existence of motion in the time-piece is originated, continued, and perfected from the philosophical principle, by which alone it acts.'

The first reference to this piece is contained in the commonplace book of the Scottish astronomer James Ferguson in 1769. His testimony as to the ingenuity of Cox's clock was quoted in the catalogue and reads as follows:

'I have seen and examined the above-described clock, which is kept constantly going by the rising and falling of the quick-silver in a most extraordinary barometer; and there is no danger of its ever failing to go; for there is always such a quantity of moving power accumulated, as would keep the clock going for a year, even if the barometer should be taken quite away from it. And indeed, on examining the whole contrivance and construction, I must with truth say that it is the most ingenious piece of mechanism I ever saw in my life.—
James Ferguson, Bolt-court, Fleet-street, Jan. 28, 1774'

The Annual Register for 1774 (vol. 17, p. 248) prefaces a quotation of Ferguson's testimony with the following words:

'Among other great works now introduced at Mr. Cox's Museum is an immense barometer, of so extraordinary a construction that by it the long sought for, and in all likelihood the only perpetual motion that ever will be discovered is obtained ...'

All the pieces in Cox's museum were disposed of by lottery, among these being the Perpetual Motion, as it was called by its maker. It passed, along with a number of other items from the collection, into the hands of Thomas Weeks who opened a museum at 3 and 4 Tichborne Street and established 'Weeks' Mechanical Museum'. Weeks allegedly carried out various improvements to the Perpetual Motion: the addition of various case ornaments which do not appear in the original engraving printed in Cox's catalogue appear to date from this time. What else Weeks did is not certain, but it is certain that here the piece ticked away to the sustained astonishment of all who paid to see it for many years. An inscription on the clock reads: 'Weeks Royal Museum, 1806'.

James Cox died in 1788. Thomas Weeks died during the early 1830s, there was a sale of many of the items at two auctions held in July and September of 1834, and here we have already lost sight of the Perpetual Motion which was not included in the catalogue. A description of Weeks and his museum of automata, which included the famous Silver Swan, is again to be found in my book *Clockwork Music* along with reproduced ephemera of the Cox and Weeks era.

Before tracing the subsequent history of the Perpetual Motion, how did it work? William Nicholson contributed a carefully written paper in 1799 to the *Philosophical Journal* which provided the basis of its operation and also, incidentally, tells us that part of the mechanism was made by a man called Rehe. His paper,[1] part of which I quote here, followed an earlier, general contribution on the attempts to seek purely mechanical perpetual motion. The writer emerges as an upholder of the quest for perpetual motion.

'In a former communication, I have given an account of some of the delusive projects for obtaining a perpetual motion from an invariable power.[2] In that paper, I remarked that the flow of rivers, the vicissitudes of tides, the variations of winds, the thermometrical expansions of solids and fluids, the rise and fall of the mercury in the barometer, the hygrometric changes in organised remains, and every other of those mutations which never fail to take place around us, may be applied as first movers to mills, clocks, and other engines, and keep them going till worn out. Many instances of this kind of perpetual motion are seen in water-mills and other common engines, which are necessarily confined to certain local situations.'

[1] 'Concerning those Perpetual Motions which are producible in Machines by the Rise and Fall of the Barometer or the Thermometrical Variations in the Dimensions of Bodies.'

[2] *Philosophical Journal*, vol. 1, 1799, p. 375.

And so Mr Nicholson went on with a rather ponderous description made all the more readable by the free use of interesting spelling such as 'pullies' and 'basons'. The reference to the engineer Rehe is qualified by the words 'who contrived and made it', concerning the barometric winding system. A footnote tells us that by June 1799 Rehe had become one of the Board of Inspection of Naval Works at the Admiralty, so we might assume that he was a respected man of proven abilities.

Fig. 73 here is based on Ferguson's illustration which must have been prepared prior to the lottery which broke up Cox's Museum about 1776. Ferguson died in November of that year.

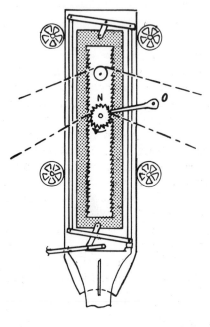
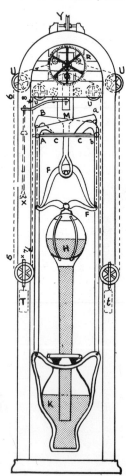

Fig. 73.

James Cox's timepiece was rendered self-winding by the attachment to it of a barometer arranged to actuate a winding wheel in such a manner that whether the mercury rose or fell, the wheel always revolved in the same direction, and thus kept the weights which supplied the motive power for the clock always wound up. Slight as the changes of atmospheric pressure may be at certain times, Cox's difficulty was not by any means their insufficiency but, on the contrary, their over-sufficiency for his purpose, and the most troublesome problem which he had to contend with was the prevention of the fracture of the chain due to overwinding. He finally overcame this by incorporating a device which caused the winding wheel to throw itself out of mesh when the weights were nearly fully wound up, and only to resume its function after they had descended again a certain distance.

The clock itself was made in a very durable manner and was jewelled at every possible bearing, the whole being enclosed in a dust-tight glass case so that the whole mechanism was clearly visible. Indeed, one feature which commended Cox's work to all who saw it was that unlike so many other engineers and craftsmen who sought to make their mechanisms obscure, Cox allowed all and sundry to see just how his perpetual motion worked.

The illustration shows a general view of the mechanism and the detail sketch shows a more detailed diagram of the barometric winding mechanism. The key to the operation lies in the two metal rocking arms, Aa and Bb. From the ends of these is suspended by rods the gimbal-mounted frame F, with its attached barometer bulb H, the tube of which goes down into the glass cistern of mercury, K. The cistern is also suspended from the two rocking arms by rods, and the method of arranging this support is the vital part of the invention. The rods supporting the bulb hang from A (the extremity of one rocker) and b (the opposite extremity of the other), while the cistern is suspended from the other, opposite extremities, namely B and a. It will be apparent that this arrangement allows that if the bulb is drawn up then the cistern must be lowered, and if the bulb is drawn down then the cistern must rise, since as either end of a rocker goes down the other end must rise. The cistern being open at the top, the varying pressure of the air forces more or less mercury into the bulb. If the weight of the bulb is increased in this way, it descends, and if the weight of the cistern is increased, the bulb ascends, being made lighter.

The frame F rises and falls with the bulb, and to this is attached the winding-up frame, M, shown on a larger scale in the detail sketch.

In this winding-up frame is a rectangular slot internally toothed rack-fashion on each side. On one side the teeth point downwards and on the other upwards. When the frame descends, the downward-pointing teeth engage the wheel N, and when it rises, the other teeth act on the same wheel which in either case is uniformly turned in a clockwise direction. The frame itself moves between four friction wheels which maintain it in an upright position. A catch, O, is provided which ensures that the wheel is always turned clockwise. At the back of the wheel N is a pulley or sprocket wheel to take an endless chain. This chain passes over pulleys UU, under the two lower ones S and s, then over the pulleys VV, and over the axis of the large wheel R by which the movement of the clock is effected by means of the weight T. The corresponding weight on the other side, t, is merely a light counterpoise to keep that portion of the chain extended. This one is just an empty brass box, but that on the other side, T, is a similar box filled with lead. Thus T acts with half its force of gravity on the part of the chain marked 5 and 6, and with half on the part marked 7 and 8.

The train of movement is such that the large wheel R would keep the clock going for a whole year before the weight T would descend to the bottom of the case. However, the alternating pressures of the atmosphere operating the winding mechanism via the toothed frame ensure that the weight is always wound up.

To avoid overwinding, Cox made the wind-up wheel so that it would turn freely on its arbor, engagement, and hence winding, only being possible when a click or pawl dropped into place on a ratchet wheel. Lifting the pawl from the ratchet and so disengaging the winding wheel was effected whenever the top of the pulley frame S reached the rod X. Delicate balance was obviously necessary at all stages of the operation and so the weight of the wind-up frame (M) itself was counterbalanced with a chain and weight passing over the pulley Y at the top of the mechanism.

As the weight T is allowed four feet of descent from top to bottom, some irregularity would have been produced by the variation in length of so much chain since the clock has a balance and not a long pendulum. To avoid this, the weight T was itself made to wind up a smaller weight every 12 hours by means of a remontoir, and it was this smaller weight, acting directly upon the time-piece, which kept it in motion.

So stood the Perpetual Motion at Weeks' Museum, a large and heavy mahogany case containing 150 lb. of mercury, a heavily founded brass and iron mechanism with heavy glass containers, and a weighty timepiece. The whole object, in short, was heavy. Added to which, it would have proved impossible to move without removing the mercury from the flasks, so stopping the perpetual winding mechanism. No doubt this is how Weeks had to move it from Spring Garden originally.

After the death of Weeks the items were dispersed, and by the 1850s the former Tichborne Street museum had been pulled down to make way for the first London Pavilion music hall. At whatever point the Perpetual Motion was removed from the building, it must of necessity have been stopped.

Its subsequent whereabouts and history are somewhat obscure. Charles E. Benham wrote an account of the clock in the *Scientific American Supplement*,[1] ending with the following paragraph:

'Almost the strangest part of the whole story of Cox's clock is the dramatic sequel of its fate, which was as regrettable as it was astonishing, and as comical as it was pathetic. The circumstances are incidentally brought to light in a work entitled "Travels in China", published in 1804 by John Barrow, who was private secretary to Earl Macartney. From this volume it appears that in the list of presents carried "by the late Dutch Ambassador" were two grand pieces of mechanism from the Cox Museum, one of which appears to have been the perpetual clock. In the course of the long journey of the Dutch Embassy from Canton to Pekin, both the machines had suffered some slight damage, and an endeavor was made at Pekin to have them repaired; but on leaving the capital it was discovered that the wily Chinese Prime Minister, Hotchang-tong, had substituted two other clocks of very inferior work-manship, and had reserved the two masterpieces for himself—it was believed with the idea of presenting them at some future time as gifts of his own to the Emperor of China, so as to gain imperial favour. Whether the perpetual clock is still ticking quietly on the wall of some chamber in the Imperial Palace at Pekin, or whether it has come home with other loot, and now reposes in some

[1]No. 1751, vol. 68, 2 October 1909, p. 212.

storehouse of second-hand goods in the east end of London, it
is impossible to say. It is at any rate satisfactory that so complete
a record of the mechanism remains, so that if anyone would
spare the money and the trouble it could be reconstructed almost
exactly as it stood in Cox's Museum more than one hundred and
thirty years ago.'

After that, it may come as something of a surprise to learn that
it is unlikely that the Perpetual Motion ever left the shores of Britain. The earliest reference to Weeks' Museum dates from 1802
and from the inscription on the clock it was probable that the Perpetual Motion was not exhibited by him until 1806. The museum
seems to have remained extant until 1837. The Perpetual Motion
then disappeared from public view and was not seen again for sixty
years. Then it appeared on exhibition at the Clerkenwell Institute
in 1898 when it was shown by its then owner, W. F. B. Massey-
Mainwaring. In 1921 it was purchased by a Mr R. G. Carruthers
in Edinburgh and placed by him on loan in the Laing Gallery,
Newcastle-on-Tyne. It was then sold by auction in London and
purchased by Messrs Blairman. The piece was acquired for the
nation in 1961 by the Victoria & Albert Museum (with aid from
the National Art-Collections Fund) where it is to be seen now,
stationary and mute in one of the public galleries. At the time when
I located it several years ago, it did not even bear a label on it
to say what it was. Yet this was and still is generally recognised
as the first successful attempt to produce a perpetual motion clock
in England (this is endorsed by the descriptive notes in the
archives of the Victoria & Albert Museum).

'The clock is referred to in the principal works on horology and
can be fairly described as one of the landmarks in the history of
English horology', comments the Museum's notes.

Alan H. Lloyd, who features it in his book *Some Outstanding
Clocks over 700 years*, states that it precedes by some 180 years
the atomic clock of today. Apart from its horological importance,
the clock is a most handsome piece of furniture and is designed
to be viewed from all sides.

Such is the change in public tastes that it is today but a minor
curio, captivating to none but the *cognoscente* of eighteenth-
century craftsmanship, clockwork, mechanics and achievement.

How can one evaluate in terms of perpetual motion a device

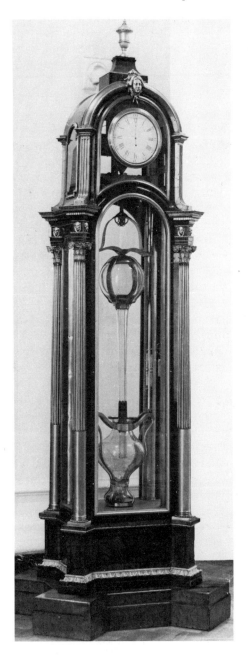

Fig. 74
Cox's Perpetual Motion.

which survives today to bear motionless witness to failure to prac-
tise its own justification? I believe the answer is to be obtained
from an analysis of what Cox set out to achieve, the manner in
which he realised his objective and the evidence which is with us
today.

First, let us examine the principle of barometric power. It was
Galileo who stumbled across the phenomenon of the pressure of
air affecting the height to which a liquid could be drawn by a
vacuum. However, Galileo failed to pursue the discovery further
and it fell to his pupil, Torricelli, to 'invent' the barometer in 1643.

Various fluids were used by perfectors during the years which
followed. Since mercury (quicksilver) is 13·568 times the weight
of water and by far the heaviest available fluid, a mercury barometer could be made much smaller and more compact than one
utilising any other fluid. It was soon discovered that in a common
barometer, the rise and fall of the mercury does not exceed three
inches. The larger the surface area of the mercury in the reservoir
(in the case of the Cox mechanism), the more energy available
to do work. This is more readily understood if you imagine
exhausting the air from a narrow-bore tube, and then containing
the vacuum easily with the finger-tip, or, conversely, retaining a
pressure in a similar tube also with the finger. If the volumetric
area of either the pressure container or vacuum retainer is comparatively large, a greater effort is needed to control it. The
pressure per square inch, or the negative pressure, remains the
same in either case. It was for this reason that the barometer which
formed the driving force behind Cox's Perpetual Motion had to
contain 150 lb. of mercury instead of just the small quantity commonly used in a mercury barometer. Had the aneroid barometer
been available to Cox (it was invented by Vidi and patented in
England in 1844), a much simpler and less cumbersome mechanism would have been possible.

A well-made barometer will, unless it is subjected to an external
confrontation foreign to its intended use, i.e. if it is dropped or
accidentally smashed, operate for what one might reasonably describe as 'forever', since there is nothing to wear out. So much
for Cox's prime mover. What about his clock?

This is a perfectly ordinary type of timepiece as regards appearance. In construction, there are some important differences. It is
robustly made so as to minimise the rate of wear and tear and at

the same time to allow it to continue to function with an eventual and acceptable degree of wear. Every possible bearing is jewelled and the clock is intended to go without lubrication. The dust-tight glass case adds credence to this design feature since it is dust which is a prime cause of the deterioration of clockwork. The minute but nevertheless present abrasive character of dust accelerates wear and slows down functioning.

The possible weak spot in the system is the winding, but here again Cox seems to have triumphed. The clock is driven by a weight and a chain. This in turn winds the prime mover of the clock which is a smaller weight.

Considering this system and the care with which it was designed, and by examining the clock as it stands today, there is every likelihood that the life of the mechanism might considerably exceed that of the oldest church clock in existence today, the difference being that such an age might be attained without maintenance and whilst deprived of the aegis of human interference in any way.

The failure of Cox's Perpetual Motion to remain in motion lies not so much in the concept of the design as in its realisation, for had the Perpetual Motion been allowed to stand in Tichborne Street from the early 1800s to today, we would probably find it still in motion and good for another few hundred years. The removal of the device proved its downfall. It was probably brought to a standstill well over a hundred years ago. It is a tantalising thought that, were it not for the extremely high cost today of the substance which once was in ready and cheap supply, this curious mechanism might again be set in going order to tick through aeons of time.

The general opinion of horologists today is, perhaps understandably, hesitant regarding the ability of the clock to demonstrate perpetual motion. But the system is all there and, however much one likes to try to discount it, and however much we may accept the impossibility of mechanical perpetual motion, this machine is a marked exception. One eminent horologist whose respected office in one of our leading museums prevented him from making statements which might be open to controversy, spent an evening with me discussing perpetual motion in its various forms. He ended by saying, 'Without any shadow of doubt, perpetual motion is quite impossible. James Cox succeeded, though, and I would dearly love to see his clock working again to prove me wrong.' Another noted authority on the same subject professed that Harrison (who invented the chronometer) and Cox deserved equal place in the annals of horology, and went on to say: 'The greatest horological paradox of all time stands in the Victoria & Albert Museum. It should be put in working order as a task of national importance.' Yet a third described the thing as a hoax but confessed that he did not know how it was supposed to work.

Was Cox the only person to attempt to harness atmospheric energy? Although probably the most successful, Cox was by no means the first to appreciate that there were forces about us in our everyday existence which might be used as a never-ending reservoir of energy. Not only was consideration given to the dilatation of matter under the influence of variations in atmospheric pressure, but humidity, dryness and temperature were each considered candidates for transforming energy into power. The Abbé John of Hautefeuille (1647–1724), who was one of Huyghens' rivals, thought up such a mechanism. This was described by the Reverend Father Dom Allexandre in his *Traité général des Horloges* published in 1736 in which he refers to the Abbé's book of 1678 containing the suggestion ·

'... a means to provide that the weight of the pendulum should be wound up by the guidance of several pinewood boards, placed transversely in two slideways, the said boards continually rising and falling according to the humidity or dryness of the air'.

Dom Allexandre added the comment that 'this invention has not been as successful as the author hoped; it has remained useless'.

Another mechanism is mentioned by Antide Janvier (1751–1835) in his *Journal Encyclopaedique* where he writes of a clock made by Kratzenstein who was a member of the St Petersburg Academy of Sciences. This clock, which was known as early as 1751, was said by Janvier to wind itself 'by alternating cold and heat', power being provided by a thermal motor.

About the same period as the Cox clock was on exhibition in London, a perpetual clock was on show in Paris. However, unlike Cox's this one was thermometrically driven by the expansion of a silver rod with the diurnal rise of temperature. Apparently seven or eight degrees of difference in temperature per day sufficed to keep the clock fully wound.

According to a correspondent of the *English Mechanic* for 1 February 1901, a Mr Burton used such a system to drive a long case clock. The writer, a Clerkenwell watchmaker, relates:

'It is run by no less a power than the solar system itself, and approaches as nearly to the idea of perpetual motion as anything I know. Briefly, Mr. Burton availed himself in the following manner of the well-known properties of heat and cold to expand and contract air. Upon the average there is a difference of 20° in the night and day temperatures. Mr. Burton placed a tin tank 10ft. high by 9in. in diameter upon a sunny wall of his house. From this tank, which is air-tight, a tube runs into a cylindrical reservoir in his cellar. In this reservoir there is a piston, the rod of which moves with a ratchet between the chain on which the piston depends. This ratchet winds the clock (an old Grandfather pattern) in the following manner. The expanding heat of the day acts upon the outside tank, and forces part of the contained air into the reservoir below. Simultaneously the piston is forced upwards. With the approach of night, however, the air in the outside tank rapidly cools and shrinks; the air in the reservoir once more ascends before the weight of the piston, and the ratchet is again put in motion. This ratchet controls the old-fashioned weights by means of which the clock is wound and the clock will consequently wind and rewind so long as the heat and cold endure and its bearings refrain from wearing out.'

A successful form of automatic power, achievable within certain limitations, was used in the clock driven by an earth battery. This method was described in Chapter 5.

Edward J. Wood in his *Curiosities of Clocks & Watches*, first published in 1866, mentions two perpetual motion clocks but does not say how they operated. One, he relates, was built in 1858 by a watchmaker named Chenhall of Drake Street, Plymouth. This he exhibited in the window of his shop and Wood said that it was about the size of an ordinary eight-day clock, 'with a novel and very simple movement, which was said to be capable of going as long as the durability of the materials permitted, without the aid of weight or spring, and in short without any manual assistance whatever'.

The second clock Wood describes was built in 1859 by James White of Wickham Market. This was a self-winding clock 'which determined the time with unfailing accuracy, continuing a constant motion by itself, never requiring to be wound up, and being capable of perpetuating its movements so long as its component parts should last'.

As regards normal clocks intended to go for a long time, Thomas Tompion, the celebrated London clockmaker, was once said to be making a clock for St Paul's Cathedral which would go for a hundred years without winding. The clock, supposed to cost between £3,000 and £4,000, was apparently never completed. A clock with a similar running period between windings was said to have been owned by the Marquis of Bute at Luton Park, Bedford. Presumably this clock was lost when the great mansion burned down in 1843. Sir John Moore wrote an account in the *Mathematical Compendium* concerning his 'large sphere-going clockwork' where we read that it made one revolution in 17,100 years, apparently through the ægis of six wheels and five pinions.

In 1934, a clock was completed in Sweden which may still be going: it was certainly still going on its initial winding a few years back. Rather like Cox's Perpetual Motion, the clock is powered by a descending weight releasing its potential energy. The interesting part is the winding mechanism. The case contains seven closed aneroid capsules of the type used in an aneroid barometer. These are filled with air, interconnected and sealed, with the lower one connected to the case and the upper one to the winding ratchet mechanism. Changes in atmospheric pressure or temperature cause the expansion or contraction of the aneroids: this operates the ratchet and winds up the weight. It is estimated that four times as much energy is supplied to the mechanism as is necessary to keep it operating. When fully wound it will run for eighteen months even without the barometric winding system. This eighteen-month period is achieved by using a torsional pendulum consisting of a heavy metal disc supported by a slender steel ribbon. Its period is seven-and-a-half seconds, equivalent to that of a swinging pendulum 187 feet in length.

Thomas Bedwell, the English mathematician and engineer who died in 1595, left notes of the many things to which he had given thought. Rather like the Marquis of Worcester's propositions, Bedwell left us with an idea for perpetual motion. He thought up a water-driven clock or clepsydra which would go continually 'without setting'. He did not say how he intended to make it work, but it obviously didn't. The time, one might suggest, was not right.

8

The Redheffer Perpetual Motion

It is interesting to observe the part which Philadelphia plays in the history of perpetual motion in the United States. This city, a major social and industrial hub since its foundation, was the breeding ground for several notable perpetual motion machines and today the Franklin Institute in that city displays not one but two attempts at seeking this rare goal. Both represent enterprises which fraudulently converted a great deal of money into the pockets of their 'inventors'.

To relate the complex story of the elder of these we must go back to 1812, the year in which Charles Redheffer (sometimes spelled Redhoeffer and in one reference Readhefer) first appeared in Philadelphia with a strange machine. He set it up in his home on the then outskirts of the city and charged admission to the public to come and watch it run. In those far-off days this sort of thing excited a great response from the public at large who had a sense of wonderment that has been grilled out of them in the intervening generations.

Redheffer's machine always ran and he never had need to touch it. As might be expected, Philadelphia was soon a hotbed of excitement. Was this really the long-sought perpetual motion? Huge wagers were made as to its authenticity. One Charles Gobert, a civil engineer who, I cannot help feeling, was probably in the pay of Redheffer, inserted the following announcement in the *Gazette* published in Philadelphia on 12 July 1813:

'I hereby offer, on demand, any bet or bets from 6,000 to 100,000 dollars, to the end of proving, in a few days, both by mathematical data and three several experiments, to the satisfaction of enlightened judges, chosen by me very opponents out of the most respectable gentlemen of this city, or of New York, that Mr. Redheffer's discovery is genuine, and that it is incontestibly a perpetual self-moving principle ... This is to be valid until the 15th inst., at sunsetting.'

Redheffer had already applied to the legislature of Pennsylvania for a grant of funds to carry out his great invention of perpetual motion and so a committee of experts was appointed to examine the project and report on its viability.

The machine to be examined was set up in a building near the banks of the Schuylkill River in Philadelphia, and on 21 January 1813 the eight appointed commissioners went to inspect the apparatus. One of them, Nathan Sellers, took with him his son, later to be the father of Professor Coleman Sellers who subsequently recorded much of the history of these events.

When the team arrived at the house, they found that the door of the room containing the machine was locked and the key, conveniently, missing, so that their only examination of the celebrated contrivance was confined to an inspection through a barred window.

The youthful Sellers studied the machine as thoroughly as he could in view of the difficulties—and made a strange discovery. The machine had a set of teeth around the periphery of its rotating table which meshed with another wheel whose axis was supposed to transmit the power to some other point where work was to be done. Sellers Jr. detected that the faces of the two wheels were polished by wear *on the wrong sides*. The sketch Fig. 75 will explain just how significant this was. If A is the rotating table of the perpetual motion apparatus driving the wheel B in the direction of

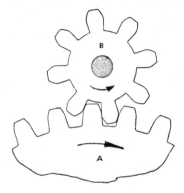

Fig. 75.

the arrows, then clearly the front faces of the teeth of A will press against the rear faces of the teeth of B, and these faces will eventually become polished by friction. If, on the other hand, it is the front faces of the teeth of A and the rear faces of those on wheel B that show signs of polishing, then it is clear that B must be driving A. This is just what young Sellers noticed and pointed out to his father as proof that the perpetual motion machine was being driven by some concealed source of power.

Nathan Sellers, satisfied that the enterprise was a fraud, took it into his head that the interests of others might best be satisfied by a sort of homeopathic object lesson. He therefore consulted with Isiah Lukens.

Now Lukens was to become one of the first two elected vice-presidents of the Franklin Institute and first chairman of its Committee on Science and the Arts, and he gained these distinctions following the formation of the Institute in 1824 on account of his great talents as an engineer and experimenter. Sellers described the machine he had seen and commissioned Lukens to make an almost exact replica operated in the manner which young Sellers had detected.

Lukens discharged his mandate extremely well, and his model is illustrated here (Fig. 76). It consists of a horizontal circular table, attached to and supported by a central vertical shaft, resting on a pivot below and steadied by a journal held in the framework above. Two inclined planes, mounted on wheels like little pit trucks, rest on this circular table and each inclined plane has on it a small truck containing two removable weights. The inclined planes, and also the trucks, are attached to levers which are supposed to transmit to the central shaft the tendencies of the inclined planes to run from under the cars and of the cars to run down the inclined planes, and these tendencies are supposed to cause the rotation of the central shaft carrying with it the circular table and all the parts on it.

The model works beautifully: if the weights are taken out of the little trucks the thing comes to rest, and the moment they are replaced it starts up again.

But the apparently solid baseboard of the model is built up from pieces of thin wood and conceals a hollow centre within which is a cleverly made clockwork motor of appreciable power and slender thickness. The whole 'perpetual motion machine' is covered by a glass case with four ornaments on top. One of these knobs is the winder and an attendant can keep the machine wound daily by the simple pretence of polishing the case. Motion from the clockwork motor drives a small plate on which rests the pivot of the central vertical shaft and the friction of the components is arranged so that when the weights are removed from the little trucks, the friction between pivot spindle and plate will be insufficient to transmit the continual motion of the motor, but as soon as the weights are put back, sufficient friction is restored and the plate turns the shaft.

Sellers and Lukens then arranged a demonstration of their model and made sure that Redheffer was invited to attend. Redheffer was so amazed to see the machine at work, apparently

Fig. 76. Model of the Redheffer perpetual motion machine in the Franklin Institute, Philadelphia. *Pictures by courtesy of the Franklin Institute.*

demonstrating the very principle which his had failed to perfect, that he privately offered Sellers a large sum of money if he would reveal to him 'how it was done'. Sellers did not at that time reveal the *modus operandi*, but made sure that all were informed that Redheffer was a charlatan.

Redheffer had been able to sap hundreds of dollars from Philadelphians and although his welcome in that city was obviously outstayed, he entertained no doubts at all that he could repeat the show in New York. Communications at that time being primitive, he assumed, correctly as it turned out, that New Yorkers would generally have no knowledge of his trouncing in Philadelphia. He opened his exhibition of perpetual motion in New York in 1813.

If Nathan Sellers and his son had, with the aid of Isiah Lukens, proved his downfall in Philadelphia, he was now in for a much rougher time in New York at the hands of Robert Fulton, the talented mechanical engineer (1765–1815). Fulton's part in unmasking Redheffer is recounted in *The Life of Robert Fulton* by Cadwallader D. Colden, published in 1817.

Fulton, it seems, was a disbeliever in Redheffer's claims, and although hundreds were daily flocking to pay their dollar to view the wonder, Fulton would not allow himself to follow the crowd. After a while, though, Fulton was persuaded by a number of his friends to visit the machine which was displayed in an isolated house in the city suburbs.   Soon after Fulton entered the room where the device was exhibited, he exclaimed, 'Why, this is a crank motion!' He had noticed that the machine must be being moved by a crank manually operated since this is extremely difficult to operate at an even speed and therefore the velocity is unequal during a revolution. A practised ear may also detect that the sound of the machine varies with its speed of motion. Had the machine been all that it claimed, Fulton reasoned, its movement and its sound would have been regular.

Fulton questioned the exhibitor—Redheffer—and then openly denounced him as an imposter. Redheffer became angry and blustering but Fulton maintained his assertion and said that at the risk of paying any penalty if he was in the wrong, he would attempt then and there to prove it.

With the approval of all those present, excepting, no doubt, the exhibitor, Fulton began knocking away some pieces of thin wood which appeared to be no part of the mechanism but to pass between the frame of the machine and the wall of the room and serve simply as supports to steady the device. These strips of wood concealed a catgut belt drive which passed through one of the pieces of wood and the frame of the machine to the head of the upright shaft of a principal wheel. The other end of the drive passed through the wall and along the floors of the second storey to a back attic some yards from the room where the 'perpetual motion' was to be seen. Here Fulton found the real source of the power which turned Redheffer's machine—an old man with a long beard who displayed all the signs of having been imprisoned in the room for a long, long time. The man had no notion what was happening and sat there on a stool gnawing a crust with one hand and turning the crank with the other.

The crowd turned mob, demolished the perpetual motion machine and Redheffer fled, a stop having at last been put to his deception which had earned him a tidy fortune in Philadelphia and New York. In Scharf and Westcott's *History of Philadelphia* there is a reproduction of the advertising circular which Redheffer used to attract visitors to his exhibition at Germantown, Philadelphia. This reveals that the charlatan had the audacity to charge $5 admission for men, but allowed women to be admitted *gratis*.

Professor Coleman Sellers, writing in *Cassier's Magazine* in January 1895, makes the interesting revelation that Isiah Lukens built not one but two models of supposed perpetual motion machines on the 'principle' of Redheffer. He identifies these as being one built for the Franklin Institute, and one of later construction built for the Philadelphia Museum collection which was catalogued not as 'Redheffer's' but as 'Isiah Lukens' Perpetual Motion'. Coleman Sellers wrote:

'I have no information regarding this latter model other than that afforded me by one of my brothers, now nearly ninety years of age, who, when much younger, explained to me in what particulars it differed from the model now existing. The museum model was probably lost with that part of the Peale collection which was destroyed by fire in the building at the corner of Seventh and Chestnut streets, Philadelphia, where it was then stored.[1] The· late Mr. Coleman Sellers, the writer's father, who died in 1834,

[1] This was the Chinese Museum which also housed the celebrated fake automaton chess-player of Von Kempelen (described in *Clockwork Music*). The fire occurred on 5 July 1854.

was intimate with Isiah Lukens, and the changes in the museum model, it is said, were made at his suggestion, and, doubtless, under his supervision, in the interest of his father-in-law, Mr. Peale, for whom it was built.

'In the Franklin Institute model, the clockwork that gives motion to the wheel is hidden in the baseboard of the machine and probably has not been seen by anybody for very many years. It is wound up by turning one of the four knobs on the columns that support the frame of the machine. The power is conveyed to the main wheel, upon which the inclined planes and wagons are supported, by means of a vertical shaft, with a spur wheel gearing into a second wheel upon the main shaft, these two wheels being entirely concealed in a brass yoke, beneath which yoke there is room to pass a strip of glass about as wide as an ordinary microscope slide. A similar glass slide seems to support the lower step of the driven shaft. These pieces of glass can be removed, and apparently the vertical shaft of the main wheel and the shaft carrying the driven pinion rest upon this glass, being steadied only by the yoke, but, in point of fact, the glass has nothing to do with the mechanism.

'In the museum model this part of the apparatus was entirely different. The spindle of the machine passed into the base-board, apparently surrounded by a glass bushing and resting upon a glass plate at the bottom. The step had the appearance of a stationary glass support for the step of the upright which was of polished steel. When the clockwork was in operation this glass plate was continually revolving, but it required the weights to be placed upon the carriages on the inclined plane to give sufficient adhesion to cause the wheel to revolve. The steel step was not ground perfectly true on the base, but touched on one side only, in such a manner as to make it act as a crank through the slight adhesion to that part of the glass which was roughened enough to give it the adhesion it required. The museum model could be taken apart and all the separate parts laid upon the table and examined critically by any curious person, and when reassembled and the weights put on, the motion would begin, at first slowly, and at last acquiring the full rate that was capable of being accomplished by the clockwork. It was this peculiarity of taking the machine apart and putting it together again that gave the opportunity to wind up the clockwork, as in securing the knobs on the posts to fasten it together, this motion served to wind up the clockwork for the time being.'

In Colden's description of the destruction of Redheffer's New York machine, he says that the catgut string led through one of the laths and the frame of the machine to the upright shaft of 'a principal wheel'. There were but two wheels in the device, the large one and a lantern pinion on an upright shaft into which it meshed. In the Philadelphia exhibition, according to Sellers, this shaft ostensibly gave motion to a grindstone or some other working machine, which device was in reality driving the perpetual motion. This was proved by the way in which the pins in the lantern pinion on the upright shaft pressed upon or marked the teeth of the large wheel. Insufficient as the description is in *The Life of Fulton*, it substantially agrees with the supposed conditions that were explained by the disbelievers who witnessed the invention in Philadelphia.

Sellers made it his business to locate people who had direct recollections of the Germantown exhibition and also to assess the qualifications of those men who formed the team appointed by the State legislature. One of them was Oliver Evans, one of the most noted engineers of the day, and all were unlikely to believe it to be anything other than a fraud. Their skills and knowledge would have been such as to prevent their being deceived by anything that was at variance with the laws of Nature.

From all the descriptions it appears that the mechanism as demonstrated by Redheffer in Philadelphia was operated in a different manner from that which he showed in New York. In Philadelphia a grindstone was operated by the machine and apparently there was power enough for steel instruments to be sharpened on it. One shaft of the grindstone approached suspiciously close to the partition of the room and was hidden by some pieces of wood that had no apparent connection with the machine. It was thought at the time that an extension of this shaft passed into an adjoining room and there a man was concealed who operated the whole thing by turning a crank.

The day after the Philadelphia City Council passed a resolution appointing a committee to ascertain whether a machine on Redheffer's principal might be capable of raising to a sufficient height a sufficient quantity of water for the use of the citizens of the city, there appeared an article in the *Aurora*. This piece, signed 'Rittenhouse', stated that the machine had never been seen in actual operation for more than half a day, and claimed that it was a deception; and it was at this point that Lukens made the model now in the Franklin Institute.

The *Aurora* subsequently became a clearing house for a battle between the believers in Redheffer's perpetual motion and its opponents. This controversy seems to suggest that the *Aurora* and its editor were believers in Mr Redheffer.

One more thing needs to be said about the model now in the Franklin Institute which is illustrated here. In many early pictures this is depicted with a model on the top representing a man crouching on top of a trunk and trying to lift himself and the trunk by the straps. This had nothing whatsoever to do with the model but was presented to the Franklin Institute and was for a while placed on top of the model as an example of the impossible.

I have chosen this point to describe another fraudulent perpetual motion machine, this one of the wheel type with overbalancing weights. It is justifiable to include it in this chapter for several reasons: it was powered by concealed clockwork, it defrauded people in another, smaller town in Pennsylvania, and its inventor was exposed. The story was related in the *Scientific American* for 1 July 1899.

In or about 1897 a gentleman named J. M. Aldrich exhibited to certain people in Bradford, Pennsylvania, a machine which he called a motor. On the strength of its apparent unique performances, he secured several not inconsiderable sums of money from the favoured few to whom he showed his model. In return, he 'gave them an interest in his invention'. However, there appeared too great a discrepancy between the promise and the performance of the motor, added to which an increasing number of people appeared to be being allowed an interest in the project in return for money. Finally, Aldrich was arrested and detained for three or four months in Auburn prison. Unfortunately for the public at large, the Bradford victims seem to have been unable to produce enough evidence to secure a conviction and he was released.

A wiser man might have learned a salutary lesson and taken up some other occupation by way of earning money. Not so Aldrich who promptly cast his net for fresh victims and continued to do so for two more years. In March 1899, however, one of his many assured half-interest holders took possession of the model and sent it to the Patent Office for evaluation. It did not take the examiners long to discover the real motion, which was not of the perpetual type but by means of a mechanically explainable spring.

The model, however, was so interesting in its design, with plenty of diversions in the way of moving parts, that experts concluded that it had set out to be a genuine perpetual motion machine, and that when Aldrich discovered that for all its double-joints, cams, slides and weights, friction still brought it to a stop in a matter of moments, he then turned to clockwork to overcome that tiny problem. And when he saw how splendid his machine looked when 'under power', he decided that here was a way of recouping his expenditure on the model, and earning a living to boot.

The device Aldrich tried to build was one in which the force of gravity was intended to supply the motive power. It took the

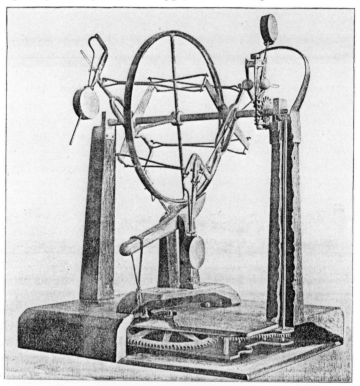

Fig. 77. Sketch of Aldrich's perpetual motion fraud showing how the clockwork motor fitted into the base.

form of a rotating shaft, two transverse arms, placed at right-angles to each other, and jointed levers which were supposed always to present an excess of turning moment on one side of the shaft. Disc-shaped weights were carried at the outer ends of two transverse arms which were themselves carried at opposite ends of the main shaft. The weights were adjusted at the ends of the swinging arms which were capable of motion through an arc of 90°. It is shown in Fig. 77.

The direction of rotation was clockwise and the weighted levers were so attached to the transverse arms that in the downward half of each revolution they fell outward and forward, thus lengthening the radius on which they travelled relative to the centre of the main shaft.

On the upward half of the revolution, the weighted levers closed up and the weights themselves described an arc of rotation with a smaller radius relative to the shaft. To assist in increasing the turning moment on the downward half of the revolution, the transverse arms were split in the centre and were capable of sliding bodily across the main shaft. As each transverse arm with its jointed lever and weight rose a little past the horizontal, it slid forward and downward, so as to throw the weight on the opposite end of the arm still further from the centre and so increase the turning moment on that side whilst at the same time decreasing that on the upward half of the revolution. The transverse arms were maintained in place by means of small rocking levers extending from steadying arms attached to the shaft.

Aldrich had made careful provision lest the acceleration and rapid motion of his machine be too great and to this end had incorporated a centrifugal governor near one of the two vertical posts carrying the main shaft. In addition, any superabundance of energy could be controlled by a small brake acting on a flywheel affixed to the centre of the shaft and held against the wheel by a rubber band.

There seems little doubt that Aldrich really did expect this ingeniously contrived machine to work, believing that his extensible arms with the weights flung far out on one side of the shaft and drawn snugly in on the other side would not only solve the quest for perpetual motion, but in a machine of suitable size would generate considerable horsepower. One can imagine the disappointment when Aldrich spun his machine round for the first

time and discovered its frowardness, for even had he been able
to eliminate friction, there still existed no turning moment.

And so its inventor tried to turn misfortune into profit by very
skilfully motorising it. The *Scientific American* was allowed access
to the machine after Aldrich had not only been exposed for the
second time, but this time also deprived of his model. An X-ray
photograph of the base told in seconds the real story, revealing
both clockwork and a high degree of artfulness in deceit. Let
the editors of that journal have the final word on Aldrich's
machine:

'The gears were connected with the main shaft by means of a small
rod extending through the right hand post, a couple of bevel
wheels at the top of the post serving to transmit the motion to
the revolving shaft and weights. The model, as it stands on our
office table, is certainly a masterpiece of deception, and eminently
calculated to deceive the unwary. The problem of concealing the
joint, after the "works" had been inserted in the hollowed out base
of the machine, was solved by forming a bevel joint and making
it coincident with the bottom edge of the base ... This has been
done so skillfully [sic] as positively to defy detection, and the
illusion is further assisted by the extreme roughness with which
the other joints on the machine have been finished. By pushing
the little block, which carries a brake, to one side, it may be lifted
away, exposing two openings in the base for winding the springs.
Considering the artistic clumsiness with which the whole affair
is put together, the worm holes neatly drilled, but drilled with
that careless abandon which marks the ravages of the native worm,
the coarse, rough jointing of the posts standing in close proximity
to the exquisitely finished bevel joints of the base, one cannot but
regret that the unquestionable dexterity of the inventor was not
directed to a better end.'

This was as clever a fraud as the clock made 120 years ago by
a man called Geiser of Tübingen, Baden, who claimed to have
perfected perpetual motion in an ingenious and simple manner.
After his death his timepiece was very carefully examined, both
inside and out, whereupon it was found that the framework supporting its cylinders was hollow and contained cleverly constructed hidden clockwork which was wound up by inserting a
key in a small hole hidden under the second hand.

Then there was Adams who for eight or nine days exhibited his pretended perpetual motion in London and relieved his admirers of a tidy sum of money. One member of his audience, though, took hold of part of the machine and lifted it a little. This seemed to disengage some hidden clockwork for he heard the familiar sound of a spring running down. The exhibitor was furious, hastily set the part back in place and continued his exhibition. Later, with a friend, the suspicious observer seated himself to one side of the machine, his partner at the other. Then, in spite of the protestations of the showman, they lifted the same wheel again, there being some play in its pivots. Immediately the hidden spring began to run down and they continued to hold it until the sound had ceased, whereupon they replaced it and offered the owner £50 if he could set it going again. Adams could not, a constable was summoned and he ended up before a magistrate and there signed a paper confessing his so-called perpetual motion mechanism to be fraudulent.

But it was a quite different solution to perpetual motion that was proposed by a Mr Gilbert, qualified electrical engineer, who

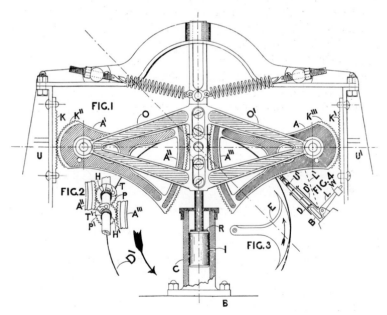

Fig. 78. Arthur E. Gilbert's perpetual motion, 1902.

announced his scheme in the pages of the *English Mechanic* for 23 May 1902. His description opened:

'It is with the firmest conviction, after twenty years' experiment and deep thought, of the possibility of a weight being made to revolve mechanically on an axle that I write this paper.

'The object of my invention is to provide a motor to work continuously, or as long as may be required, by its own mechanical parts without any external aid, for driving a dynamo, electric machine, and for other purposes, thereby dispensing with fuel. To attain this long-sought result, I utilise the natural pull or throw of a weight, or weights, which may be in the form of a disc with a knife-edge to cut the air, in connection with an axle on which may, or may not, be a flywheel.

'Under one arrangement I employ inclined uprights braced together at the top and fixed to an inclined base, the uprights inclining at an angle of 45°, more or less, in the direction in which the weights descend, but they may incline in the opposite direction...'

And so he goes on in the style of a patent specification to describe a machine which not only has 'inclined uprights' (?) but also 'goes of itself'. Cams, springs and his knife-edge flywheel provided the notion, but not the motion.

9

Keely and his Amazing Motor

Of all the perpetual motion frauds the story of John W. Keely's carefully planned deception and the manner in which the Keely Motor Company defrauded people of large sums of money must rank supreme. As a perpetual motionist, it is difficult to imagine that he ever set out with honourable intent. That he made a good living out of his 'inventions' there is no doubt. There is also no doubt that there were many who regretted that he died before the fraud could be exposed. Had this happened in his lifetime, at the very least Keely would have received a lengthy session behind bars; at the very worst, he might have been lynched.

John Worrell Keely was born in Philadelphia on 3 September 1837. He worked as a carpenter until 1872 when, he later related, a tuning-fork gave him the first hint of a new motive power which he claimed to have discovered. His claims were first brought to public notice in 1873 and while they were derided by the few as being absurd, they were popularly believed to have some foundation. This belief was strengthened by the fact that several well-known scientific gentlemen who saw some of Keely's experiments during the winter of 1873–4 were unable to discover any fraud. They were, however, prevented from examining the apparatus. During the year which followed, exhibitions were made before capitalists in Philadelphia, New York and Boston, and demonstrations were made of a powerful force which Keely persuaded them was produced by the 'disintegration' of a few drops of water.

Such exhibitions of power lost nothing of their marvellousness in the retelling and before very long imaginative newspaper reporters were whipping up enthusiasm over Keely's discovery in the pages of the popular press. So earnest, sincere and enthusiastic were these reports that many people began to doubt whether the physical laws of nature had been correctly stated in the textbooks of the time.

Keely had inspired such confidence in his claims that within a few months he was able to form the Keely Motor Company with a capital of $5,000,000. One of the most plausible arguments used in floating the company stock was that if Keely established his claims by bringing out a commercial engine using his principles, all present sources of power would be superseded by the 'etheric force' evolved from a thimbleful of water, in which case a single share of the stock would be worth a fortune to its possessor. This was a powerful incentive indeed for the potential investor.

Unlike most great inventors, John Keely had no education. He was first known as an inventor by his exhibition of a perpetual motion apparatus at premises in Philadelphia's Market Street. This was several years before the advent of his motor. As regards appearance and physique, he was a powerfully built man around six feet tall, broad shouldered, square-jawed, muscular and fond of exhibiting his great strength. He spoke very rapidly, and when explaining his experiments, obtained the confidence of the listener by an apparent freedom from anything like subterfuge. He inspired respect as an honest man who concealed no facts, although his use of language is on record as being strange. He was fond of using words out of their accepted normal meaning so that an investigator would learn nothing from any explanation of his relating to the character of his force or the means of controlling it. Pseudo-technical language impressed those who heard him tell of his 'hydro-pneumatic-pulsating-vacu-engine', 'sympathetic equilibrium' and 'etheric disintegration', even 'quadruple negative harmonics' and 'atomic triplets'.

The story of the Keely motor can be divided into two clear periods which help to provide us with some sort of understanding of the character of the claims made by Keely and the experiments by which he supported them. The first period covers his claims to the production of force by the disintegration of water. He allegedly accomplished this by the use of an apparatus which he called the 'Liberator'. Those who bought shares in the Keely Motor Company had the satisfaction of seeing a picture of this on their stock certificates.

The second period began about 1886–7 when he claimed a new discovery of a force derived from the vibration of some hitherto unknown fluid between the atoms of the illimitable ether. This second claim was brought out after the Keely Motor Company, discouraged by Keely's failure after many years to bring out any practical commercial motor, had ceased to supply him with funds for his experiments or his support.

During the first period, there was wild speculation in the stock of the company, due largely to the almost hysterical reports in many of the newspapers of what had been achieved (?) by Keely, and more especially to the fact that eminent men of scientific ability, albeit in fields other than physics, had not only endorsed Keely's claims but had also become shareholders. It is not strange, therefore, that there were many thousand stockholders in Philadelphia and other cities. Their numbers included many who could ill afford unwise speculation, such as clerks, shop-girls, widows and orphans—all people looking for the day when the increased value of their shares would make them independent.

The tide of speculative investment was checked when a few conservative newspapers, including *The Ledger* in Philadelphia, pointed out the absurdity of Keely's claims, and published the opinions of well-known physicists as to the belief that Keely was a fraud. Among these were Dr Cresson and Dr Barker whose investigations were incomplete owing to Keely's insistence that he would not have his inventions pried into by those who did not and could not understand them.

What became of the money which came to Keely as his share of the great sums raised by the sale of the stock is open to conjecture. Apparently he lived in good style, was free in the use of money, gave to charitable enterprises, purchased some diamonds 'as an investment', as he told a friend, and generally behaved as a citizen of substance in the community which was supporting him. However, the withdrawal of financial aid by the Keely Motor Company had apparently left him without resources. The public seemed to have paid scant credit to the inventor and now treated him with indifference. This was the impression gleaned from a newspaper paragraph about him read by a wealthy widow. This was in 1881–2, and the future benefactor of Keely was the widow of Bloomfield H. Moore who had made a fortune amounting to

$5,500,000 in paper manufacture. He had died in Philadelphia on 5 July 1878.

The newspaper paragraph related that the inventor, still struggling to perfect his apparatus, was on the verge of starvation and despair. Mrs Clara J. Bloomfield Moore later said that she read this news following another story about a New York inventor who had been unsuccessful in getting anybody interested in his invention which, after his death, was seen to be of great value. The benevolent Mrs Moore's heart went out to Keely for she thought that here was an opportunity to save another inventor from a similar fate. She tracked down Keely, visited him—and so began a fresh flow of cash into the Keely coffers.

Within a year or so of this rehabilitation of his fortunes, Keely announced the discovery of his vibratory force. Meanwhile, Mrs Moore had become a convert to his theories and through her able pen and her many influential contacts, not only in the United States but in Europe as well, she gave wide circulation to these theories. This revived the hopes of the Keely Motor Company shareholders. Keely, however, had other ideas. The company might press for some reimbursement from him (meaning Mrs Moore) and he did not choose to do anything to upset the Golden Goose. Keely told the company that it had no claims to his new discoveries made since the company disowned him. One shareholder put the case to the test and took Keely to law to compel him to show wherein the two discoveries differed from each other.

The court ruled that Keely should explain the difference, and when he refused to divulge the secret he was sent to prison for contempt of court. A compromise solution was found by Keely agreeing to permit a mechanical expert to examine the apparatus and make a report. The expert did his best, and found that whatever Keely believed that he had invented latterly, it was different from whatever he thought he had invented earlier. The court was satisfied, and Keely was released from jail.

Mrs Moore, who had unshakable faith in Keely's integrity, invited the leading physicists of the United States and Europe to examine and report upon Keely's discovery. The few who accepted the invitation were not allowed by Keely to handle the equipment, or to do more than remain spectators of his experiments and demonstrations. Some came away puzzled by what they saw; others formed theories of how well-known forces of nature would account for the results produced. However, nothing they could say or publish could shake the faith of either Mrs Moore or Keely's followers.

Mrs Moore was of English birth and always retained her preference for London over Philadelphia. After an absence of several years in England, she returned to the States to sort out some litigation concerning the trusteeship of her late husband's estate. Her championship of Keely's fortunes was an element in the case and she chose to fortify her position by endeavouring to get several eminent physicists to examine Keely's inventions, believing that in their apparently advanced state of perfection the experts must make favourable reports. These gentlemen, among whom were Thomas Edison and Nikola Tesla, all declined the opportunity for various reasons. Time, for all the parties concerned, was running out. Mrs Moore, still believing Keely to be sincere, wanted her protégé to serve as support for her court case. Her very endeavours to secure this marked the beginning of the end for the whole enterprise and, sadly, for the leading characters in the story.

In November 1895 Mrs Moore invited the president of the Spring Garden Institute, Addison B. Burk, to make an investigation. Burk asked if he could bring along an electrical engineer named E. Alexander Scott. The request was acceded to, and in fact Scott took charge of the investigation. This seemed natural since Scott was familiar with the history of the Keely movement and had talked with the inventor about it back in 1874. He was also familiar with the views of some United States government engineers before whom Keely had performed some experiments at Fort Lafayette.

Scott's first visit to Keely's workshop at 1422 N. Twentieth Street, was made with Mrs Moore on 9 November 1895. The afternoon was spent with Keely. On this and subsequent visits he gave Keely no impression that he did not accept the inventor's statements as fact, since any question which might suggest doubt would certainly have prevented him from seeing other experiments by which he might confirm his belief that Keely really was a trickster. His patience was rewarded and many things were shown to him including some of which he had heard tell several years earlier. Among these was a levitation experiment, by which heavy weights were made to rise and fall in water in response to musical sounds produced at certain pitches. This experiment had been shown to admiring investors and investigators from the earliest days of the Keely mythology, and had always proved very effective. Scott left the building that afternoon satisfied that the experiments he had seen did not depend on any hitherto unknown source.

On the second and third visits, Scott was accompanied by Burk. Both men afterwards analysed what they had seen and concluded that compressed air was used in nearly all the experiments either as the moving force or as auxiliary to some other force more powerful but hidden from view. As an instance of this, the rise and fall of the weights in a jar of water, closed at the top, was found to be the rise and fall of hollow globes and discs, delicately balanced, so that an increase or decrease of pressure in the enclosed air space above the water would make them sink or rise to the surface. The necessary variation in pressure was produced by the introduction or extraction of air into this space through a small tube which Keely assured them was a solid wire. During one visit, Scott picked up the end of the wire when Keely wasn't looking, and found that it was hollow. The tube was a common feature in nearly every piece of apparatus in the laboratory. Other apparatus which had so excited the admiration of many prior investigators was likewise found to operate on accepted and perfectly normal principles. Keely, the two men now confirmed, had discovered no new force—only how to delude people. They reported this to Mrs Moore, who as one might expect was somewhat surprised.

In February 1896 Mrs Moore sent for Professor W. Lascelles-Scott, an English physicist. Her reason for this was that Keely was ageing, and she still felt that he had invented something important and wanted Keely to impart the knowledge to the learned Englishman in case he should die before completing his commercial engine. The Professor was allowed to examine whatever he wanted and to have full instructions as to its use from Keely. After a month of investigation, the worthy Professor stated before a meeting at the Franklin Institute that 'Keely has demonstrated to me, in a way which is absolutely unquestionable, the existence of a force hitherto unknown.' Bold words indeed!

Mrs Moore was elated and soon afterwards arranged for the disbelieving E. A. Scott to witness experiments at the laboratory devised by Professor Lascelles-Scott to show that the former's verdict was in error. This was duly done and the two men and Mrs

Fig. 79. Top: John Worrell Keely in his laboratory. Bottom: Keely's 'generator' which turned tap water into high-pressure 'etheric vapour' when 'vibratory energy' was applied. After the death of its inventor, it was found to run on compressed air as did his 'motor'.

Moore then agreed it would only be right and proper for Keely to repeat the experiment; but this time Mrs Moore would prove whether or not the hollow wire really was a vital tube by cutting it while the experiment was in progress. Mrs Moore agreed that this would be the real test, and so on Sunday, 3 May 1896, the Professor went along to see Keely to tell him what had been decided. What happened at this interview he duly recorded in a letter to Mrs Moore. 'To my surprise', he wrote, 'he declined point-blank to repeat the demonstration of Saturday, just as I was told he would do [by E. A. Scott].' The Professor urged Mrs Moore 'the absolute necessity of winning Mr. Scott's opinion at once, in Keely's own best interests as in yours'. He also said: 'It would be rashness to discuss the subject further at the Franklin Institute.' Without calling on Mrs Moore, he promptly returned to England. Mrs Moore immediately withdrew her financial support from Keely. This was 4 May 1896.

Keely still protested the validity of his devices but, it was later revealed, Mrs Moore lost faith in him and for the next few years all she gave him was an allowance of $250 a month.

On 18 November 1898 Keely died, taking with him to the grave many of his pseudo-mechanical secrets. But he was not to be spared the final exposure, for the Keely Motor Company immediately took possession of Keely's laboratory. With the aid of a Boston electrician named T. Burton Kinraide, who had been in touch with Keely at various times, the laboratory was dismantled and the engine removed to Kinraide's home at Jamaica Plains. Here he apparently spent some while trying to get the thing going.

Meanwhile back in the Keely house, E. A. Scott and Burk began an investigation of the house itself. The Keely Motor Company had concentrated on the removal of the motor and had not bothered with the rest of the equipment. On 20 December 1898 Kinraide discovered his first evidence of fraud. Now Keely's widow, hitherto very much in the background, was in the middle of legal threats from the Keely Motor Company, and her counsel, Charles J. Hill, revealed to the world on 29 January 1899 the whole truth about Keely and his mammoth fraud, none of which it seemed the trusting widow Keely knew anything about.

The news broke in Philadelphia with justified clamour. The papers carried banner headlines 'The Keely Motor Secret is Out'. The *New York Journal* for 29 January ran a banner 'Keely, the

Monumental Fraud of the Century!' And a close friend in Phila-
delphia revealed that, when one day he said to the ageing Keely,
'John, what do you want for an epitaph?' Keely had replied,
'Keely, the greatest humbug of the nineteenth century.'

Strange to say, some of the directors of the Keely Motor

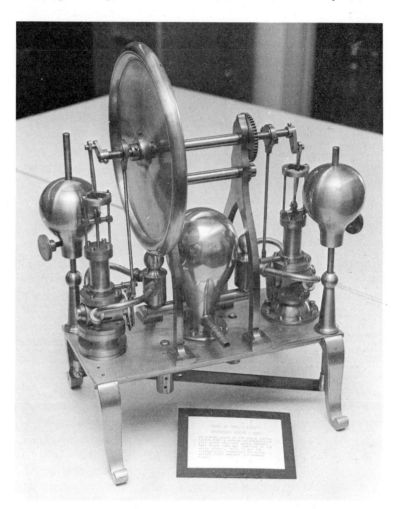

Fig. 80. A model of Keely's 'hydro-vacuo engine' of 1872, now on show
in the Franklin Institute, Philadelphia. *Pictures by courtesy of the Franklin
Institute.*

Company still had faith in the enterprise and struggled to resurrect the business for a while. Meanwhile, Kinraide went on record as believing the Keely machinery to warrant classification among the greatest frauds of the century.

As for Mrs Moore, her health had been failing and she returned to her London home before Keely's death. News of his demise was brought to her by her grandson who was Count Eugene von Rosen, attaché to the legation of the King of Sweden and Norway at the Court of St James's in London. He said that when he told his grandmother the news, she was unmoved. 'I hope', she said, 'that he imparted the secret to someone before he died.' On 5 January 1899 she too died. It was said that her life had been shortened by the final years of the Keely fiasco. She left more than $1 million excluding real estate.

What about Keely's 'strange forces'? Well, his devices, many of which were linked to musical instruments such as an autoharp, a mouth organ and tuning-forks, were claimed to be set in motion by the amplification of the interference set up between different sound waves. His motor, on one quart of water, would (Keely claimed) run a train from Philadelphia right across the continent to San Francisco, and a ship could sail from New York to San Francisco on just a gallon of the same common liquid. Terms like 'molecular vibration', 'oscillation of the atom' and countless others were all part of the humbug.

The examination showed that Keely's use of wires or thin 'rods' to convey amplitude was the key to everything. The importance of these seemingly solid rods he never quite explained, but during demonstrations he would apparently show the solidarity of them by filing one. In truth, these rods were microbore tubes for conveying compressed air. The air reservoir was a cylinder in the cellar and the compressor was a water motor. A series of rubber bulbs concealed under the floor and which yielded to foot pressure worked the whole thing. The illustration of the Keely house, published in the *New York Journal*, is reproduced here (Fig. 81).

Keely—what can we make of him? A humbug, yes, but undoubtedly a clever one. A self-taught and fairly good engineer certainly. But he was also something of a prophet. He predicted the flying machine, adding that the perfection of his engine would make such a device possible. He invented a pneumatic gun, and built all his own equipment which was no mean feat. One person

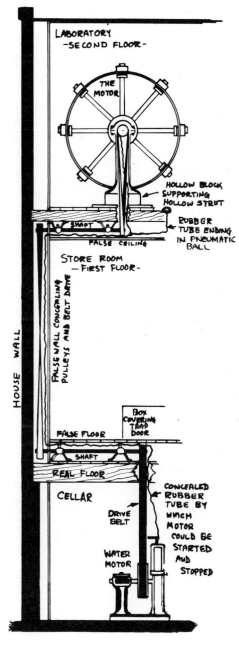

LABORATORY
—SECOND FLOOR—

THE MOTOR

HOLLOW BLOCK SUPPORTING HOLLOW STRUT

RUBBER TUBE ENDING IN PNEUMATIC BALL

SHAFT

FALSE CEILING

STORE ROOM — FIRST FLOOR —

FALSE WALL CONCEALING PULLEYS AND BELT DRIVE

HOUSE WALL

BOX COVERING TRAP DOOR

FALSE FLOOR

SHAFT

REAL FLOOR

CELLAR

DRIVE BELT

CONCEALED RUBBER TUBE BY WHICH MOTOR COULD BE STARTED AND STOPPED

WATER MOTOR

Fig. 81. The secret of Keely's house in Philadelphia. After Keely's death, the house was stripped and found to conceal the wherewithal of trickery and deception on a massive scale. When T. B. Kinraide had completed his inspection of the Keely motor he stated: 'The new force he introduced is mere bosh. I have accomplished everything that we ever saw him do, and it is extraordinarily simple. I have a magnet concealed in the wall, as he did. Then I get hydraulic pressure and compressed air from hidden sources. One of the most unique pieces of mechanism I found ... was a spring, to wind which it was necessary to use a key as big as a crowbar. With the proper winding this spring will run for three or four days and produces enormous energy. This can be wound and started running before the entrance of visitors. With the aid of these four unseen and powerful agents, Keely duped us.'

who saw it in use said that a pint of water poured into the cylinder resulted in a pressure of 50,000 lb./sq. in. being produced. No small force to be contained, however produced.

Keely did go down in history and his autobiographic epitaph stands correct. Had he not tried to capitalise on his showmanship to such an extent, he might just be remembered as an amusing exhibition of the unusual in the same way that The Invisible Girl[1] or any other of the many trickster shows of London in the last century are recalled now.

Today, Philadelphia has forgotten all about its infamous resident. Even Keely's old house has completely disappeared and the site is now a private parking lot for the property next door.

[1]Described and illustrated in *Clockwork Music*.

10

Odd Ideas about Vaporisation and Liquefaction

Any modification of the form of natural elements fascinated the perpetual motionists. Just as the alchemist sought to create something new by dissolving his metals, just as the quack doctor sought to cure the rich of their money by creating a taste, a smell and a colour, the idea of subjecting gas to such pressure that it liquefied, or heating a substance until it vaporised, became a new string to the perpetual motionists' bow. Refrigeration, heating and transformation were exciting possibilities to be employed.

The Second Law of thermodynamics meant nothing to the serious and dedicated inventor who would not permit its presence to obstruct his goal. The boundless possibilities growing out of the perpetual motion were too fascinating, its unlimited and uncomplaining response to the heightened complexity and increased demands of modern civilisation was too satisfying for it to be abandoned, and every advance in science stimulated the hope that a new principle would do away with the limitations imposed by earlier partial and imperfect knowledge.

No less an august institution than the United States Navy was involved in one strange enterprise, that of the perpetual motion machine devised by one Professor Gamgee. The Chief Engineer of the Naval Department, B. F. Isherwood, made a report to the Secretary of the Navy on this machine, extracts from which were later published in the *Kansas City Review*, volume 5, 1882 (pp. 86–9). I quote:

'From observations made by Professor Gamgee in the experimental working of this machine, he deduced the possibility of what he terms a zeromotor, in which, by means of properly adapted apparatus invented by himself, the heat in water or other objects at ordinary atmospheric temperature may be utilized to vaporize liquid ammonia under very considerable pressures, but within the control of known means of retention. The high pressure gas thus obtained being used with the greatest practicable measure of expansion on a working piston generates power, becoming by that very expansive use greatly refrigerated and diminished in bulk, and partially liquefied at the end of the stroke of the piston, when it is exhausted and then returned by a method invented by Professor Gamgee, to the ammonia boiler whence it came. The cycle is thus a closed one; no material is lost, and no heat is rejected in matter leaving the engine. The work done by the engine is due to the difference in bulk of the material when it enters and when it leaves the boiler, that difference being caused by the heat derived from water or other natural objects in the ammonia boiler and from the refrigeration resulting from the transmutation of a portion of this heat by the engine into the mechanical work performed by the latter. That this difference of bulk exists is indisputable, and if the proper mechanism can be contrived to utilize it, the idea of the zeromotor becomes realized. It will be observed that this power has not been obtained from artificial heat produced by the combustion of fuels, but from the heat of natural objects at ordinary atmospheric temperatures, and therefore costing nothing in money . . .

'The purpose of the Department in ordering an examination of Professor Gamgee's ice making machine was not to obtain an opinion on its ice making merits, but one as to whether his observations on the behavior of ammonia in the process were sufficiently accurate to warrant his inference of the practicability of constructing a successful zeromotor for industrial uses—a motor, in short, destined to supersede the steam engine. Accordingly I have closely investigated the working of the apparatus. The facts of liquid ammonia gasifying at ordinary atmospheric temperature under very high pressures, and of that gas undergoing very great refrigeration when used expansively in doing work, are not called in question by any one. Both are well known phenomena. The special fact to be observed was whether any part of the ammonia which entered the cylinder as a gas left it as a liquid, and, so far as the form of the apparatus allowed any observation to be made,

such appeared to be the case. The possibility of the invention of
a new motor of incalculable utility would seem to be established,
and in view of the immense importance of the subject to the Navy
and to mankind at large, I strongly recommend it to the serious
attention of the Department, suggesting further that whatever
facilities the Department can, in its opinion, consistently extend,
be allowed to Professor Gamgee for the continuance of his important experimental inquiries in the Washington Navy Yard. He is
most anxious to bring his invention, with the least possible delay,
to a crucial test by the completion of the necessary mechanism,
and its submission to any board of experts which may be ordered
experimentally to ascertain its merits. For this purpose he pro-
poses to use such parts of his present ice making machine as can
be re-combined in his zeromotor, adding the other necessary parts,
and thus producing, with but little loss of time, an embodiment
of his idea that will by simple trial show whether an unquestionably correct theory has been successfully reduced to practice.

'Professor Gamgee has perfected the calculations and drawings
for the mechanism required to give practical effect to his invention,
and there remains only to execute the mechanical work. He pro-
poses to use the steam cylinder of his ice-making machine as the
ammonia cylinder of the new motor, the present ammonia condenser, and the present ammonia boiler as a low pressure boiler,
adding another ammonia boiler as a high pressure boiler. These,
together with the ejector between the condenser and the low
pressure boiler, a small pump for pumping liquid ammonia from
the low pressure to the high pressure boiler, etc., will constitute
the zeromotor—a machine, as will be apparent from this brief description, of the simplest, cheapest, and most manageable kind ...

'All that remains is to give the system a practical test in order
to ascertain whether the mechanism proposed will act efficiently
enough to realize the expected result. Should this prove to be the
case, the steam engine will, within the near future, be certainly
superseded by the zeromotor, for the great item of coal, whose
cost is the principal expense of operating the former, will be
wholly eliminated with the latter. If it can once be practically
shown that a very much cheaper, lighter, and a far less bulky
mechanism than the steam engine, including for the latter its
boilers, and, in case of steam vessels, the coal bunker and its contents, can be employed for the production of power to any amount

without the use of fuel, nothing can prevent its introduction into general use for all industrial purposes ...

'The success of the zeromotor is of more importance to the Navy of the United States than to the navies of the great maritime powers of Europe with which it may come in collision, because those powers have colonies and coaling stations on the farthest shores, while the United States possesses neither, and would consequently, in naval warfare, be at great disadvantage for want of coal—its navy, as a rule, having to render service within a reasonable distance of its own coasts the sole base of supplies. If coal, however, can be dispensed with, we are at once placed on an equality in this respect, and our cruisers enabled to penetrate the remotest seas as easily as those belonging to countries having possessions there ...

'I have ventured these few remarks to show the nature and scope of Prof. Gamgee's invention, which is not that of a machine for the application of power, but for the immensely more important purpose of generating power itself, so that, strictly speaking, it includes as a basis all other machines ...'

The editor of the *Kansas City Review* saw fit to add his own rider to the foregoing. 'The invention', he wrote, 'will finally, of course, rest on its actual merits.' Gamgee's idea of using liquid ammonia in his heat engine depended on the fact that ammonia vaporises into gas at a low temperature. At $0°C$ the gas exerts a pressure equal to four times atmospheric pressure and the inventor reasoned that the transfer of heat from the surrounding atmosphere would be sufficient to transform the liquid ammonia into gas, and that this dense gas, on driving the piston in his engine and thereby expanding, would condense back into a liquid, drain into a reservoir, and start all over again.

What actually happened must have been disappointing though it seems obvious that its cause was apparent neither to Gamgee nor to those who supported him and his 'invention'. The environmental heat was no doubt ample to convert the ammonia liquid into gas, but this was nullified in the system as a whole by the cooling of the gas on expansion. Starting at $0°C$ and a pressure of four atmospheres, the temperature of the gas must drop to $-33°C$ by the time its volume has quadrupled. For the gas then to condense into a liquid, both the condenser and the reservoir

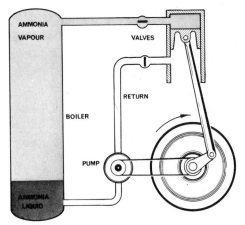

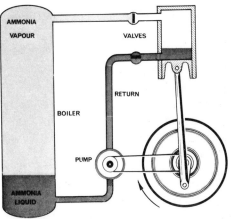

Fig. 82. The cycle of the Gamgee ammonia engine of the 1880s in which gas pressure was expected to be sufficient to drive a piston. As the gas expanded the inventor assumed that it would condense spontaneously and return as liquid to the boiler.

must be at a temperature lower than $-33°C$. No provision had been made for this cooling and, were it to have been made, the cooling process would have necessitated more energy than the zeromotor could provide. The Second Law of thermodynamics remained inviolate, and the zeromotor remained stationary. I have been unable to trace any further reference to the goings on in the Washington Naval Yard over which Isherwood enthused so much. Indeed, President Garfield and several other Cabinet members had inspected models of the Gamgee motor but, for all this, the future of the US Navy was not to depend for its continuance on the success of an ice-making plant.

Daniel W. Hering, in his *Foibles and Fallacies of Science* (p. 89), reminds us that by 1895 gases had been liquefied by the so-called regenerative method with less difficulty and expense than had hitherto been possible.

A few years later, a Mr Charles E. Tripler of New York devised apparatus for the liquefaction of air on a large scale and this resulted in a popular article concerning Tripler's laboratory and what was then considered his remarkable work appearing in *McClure's Magazine* for March 1899. This article was written by Ray Stannard Baker, a member of the editorial staff of the magazine, and contained some startling statements and one especially which implied the refutation of the Second Law of thermodynamics, and the achievement of perpetual motion (see also Chapter 13 below). Tripler was quoted as having said:

'I have actually made about ten gallons of liquid air in my liquefier by the use of about three gallons in my engine. There is, therefore, a surplusage [sic] of seven gallons that has cost me nothing, and which I can use elsewhere as power.'

The very cold liquid air in the boiler of an engine would be vaporised and have high pressure under the heating effect of the atmosphere, without any other fuel, and the air thus under pressure would drive the engine which, in turn, would compress more air to be liquefied and employed for power purposes. The use of the air for driving the engine constituted no difficulty either in theory or practice, but according to accepted ideas of science, as much work would be required in compressing the air and depriving it of heat as the air could possibly restore in again reaching its normal pressure and temperature. Still, there was Tripler's statement which he offered to verify in his laboratory. At the invitation of *McClure's Magazine*, through Baker, two professors, heads of the departments of physics and chemistry in a prominent university, visited Tripler's laboratory to witness such a demonstration. The visit, though made by appointment, proved not to be conveniently timed for Tripler, and nothing came of it except a brief comment from each of them criticising Tripler's claims. This the magazine did not publish, and the exploitation of liquid air and its wonders continued. Those who had declared war to the death on the Second Law of thermodynamics were elated and exultant.

Tripler disliked calling his invention a scheme for perpetual motion, always insisting that the heat of the atmosphere was a furnace for his liquid air, and consistently refusing to admit that he lost any power in getting the air to a temperature below that of the surrounding bodies, i.e. denying the validity of the Second Law of thermodynamics. The promises of the liquid air scheme were alluring—bewilderingly so—and its friends were loath to give up the hopes based upon them. Posing as an exemption from a painful but inexorable law, this fallacy lingered for several years and died hard.

But while on the subject of air and properties which it was thought to have, let us move forward into the present century and take a look at a pamphlet entitled 'Die Perpetuum Mobile Theorie' by one Franz Hoffmann of Saalfeld, then in Prussia. This curious piece was published in Leipzig in 1912, three years after an international aviation meeting at Rheims in which the Germans had been thoroughly trounced by the French and American contesting aircraft (for chauvinistic British readers, I should point out that there were no British entrants at this meeting held on 22 August 1909). The Great War was still two years away when Hoffmann promulgated his rather involved scheme, the gist of which can be gathered from his closing paragraphs steeped as they are in naïve patriotic sentiment:

'Any one who cannot understand that, there is no help for,—it will happen with him just as with certain gentlemen who, some ten years ago, had not been able to understand that a body that was essentially heavier than air could nevertheless lift itself free in the air. The consequence of this intellectual debility was that three years ago in Rheims we had to let Messieurs Frenchmen and Americans fly away from us instead of the Germans leading the remaining nations in flying. Perhaps a gracious fate may preserve poor Germany from another Rheims humiliation that will come from the fact that not until other nations arrive in Hamburg or Bremen with their "perpetual motion" ships, will the German Michael awake from his lethargy.'

Hoffmann must go down as a truly dedicated perpetual motionist, for he winds up imploring every reader who still has any regard for Germany's name and honour to do what he can so that, at least in respect to perpetual motion, Germany may remain in advance of the other nations!

There sometimes comes a time when a man who is master of one branch of science believes that he has found the solution to a problem involving another. The invention of one new piece of revolutionary mechanism has often engendered the making of wild claims in other directions. Some of these claims have been preposterous only by virtue of the nature of the technology ruling at the time of their being put forward. Thomas Edison visualised the day when a telephone could also transmit television pictures of the caller and the person at the other end. Technology has only recently caught up with this prophecy. H. G. Wells and Jules Verne were visionaries of a similar kind. On a broader field, science still has a long way to go to catch up with the inventive perspicacity of the science fiction writer.

It was steam which captivated the mind of a perpetual motion seeker named J. S. Hamilton of New York at the end of the last century. His idea made use of then modern technology in the form of the injector valve, except that he reversed it so that it operated as an ejector. Since an injector, by means of a steam jet, will allow a stream of water to enter a boiler against a pressure equal to or greater than that of the steam jet, then, according to Hamilton, if a stream of water flowing out of a cistern at a high level were to have its velocity sufficiently increased, it would re-enter the cistern at a lower point and would be capable of doing work in its passage external to the cistern. Hamilton wrote:

'Starting the turbine from exterior source, (motor or engine), established the vacuum [below it], after which the turbine will run alone. The initial pressure will seek the vacuum and perform work en route. The water will return by reason of its increased velocity secured by the nozzling effect of the passage ways inside the turbine. The entrance gates of a water turbine nozzle the water, and since the turbines are radial inward flow, the passage ways in the "runner" are more narrow near the center where the water leaves it. Provided the water's velocity is increased it will enter, just as the injector has proven times without number.'

The inventor has said nothing of the means to keep the turbine in motion if the water, on leaving it, is to have a greater velocity, and therefore more energy, than on entering it. It is easy to demonstrate that the successful performance of such a motor would contradict the conservation of energy. If it did not, then the thing which so many perpetual motionists feared would surely happen. The motor would accelerate at an incalculable rate as its energy mounted until it flew apart (notice the number of perpetual motion motors which were provided with a means of braking!). Such an acceleration to a virtual infinity is the basis of atom-smashing, but this is not done with steam and a boiler.

11

The Astonishing Case of the Garabed Project

Once in a while, usually when young, we suddenly believe that we have perceived something from an entirely new angle. We find we have *invented* something, or come up with an apparently earth-shaking idea. Why, we wonder, did no one ever think of that before? The cold and sober answer is that either it has been thought of and long-since discarded, or in the very clarity with which it has presented itself to us, we have been dazzled out of recognising the impossibility of it. A very senior Royal Air Force officer, both veteran aviator and qualified engineer, once confessed to me in confidence that he thought he had solved the mystery of flying at the tender age of seven. All you had to do, he reasoned, was jump into the air and, at the top of the jump, jump again, and keep on until you got to whatever height you wished. Bootlace flyers are in the same class—those who are supposed to believe that if you pull hard enough on your shoelaces you will take off!

All schoolboys have, I am sure, at some time experimented with the strange power of the flywheel and observed its apparently immense power, not to mention its behaviour as a gyroscope. This brings me to the events which took place in the United States as recently as the late autumn of 1917.

America is a mongrel country where people of all nationalities and descents rub along together. And so when an Armenian, Mr Garabed Giragossian, urged the attention of the American public for his invention of 'free energy', his plea to his fellow men was received without any form of prejudice—even the free energy bit. Giragossian suddenly came wide-eyed and enthusiastic into the public eye and made his claim with the simple conviction of one who with his discovery had surprised not just his audience, but himself as well.

Scientists and those who make great discoveries are usually discerning men who have the backing of education and a training which has adapted their mental approach to thinking and reasoning. The need for a carefully developed power of accurate observation and the ability to draw correct conclusions has long since rendered scientific discovery, once an open field for all-comers, much more the province of those who have demonstrable qualifications. Of course there are exceptions, but they are rare. So when Giragossian was confronted with the bald question 'What are your qualifications for undertaking this work?' his inability to comprehend that it was generally accepted that engineering qualifications would add credibility to his claim, caused raised eyebrows. In fact, his only answer was to assert that he was an honest man and to offer proof of this fact in the shape of glowing testimonials from his technically nondescript collection of friends and sponsors.

But Garabed Giragossian claimed that his discovery was the greatest thing of all time, and so certain was he of its worth that he wanted the United States to assume its development for the benefit of the nation as a whole. He made representations to Congress which could not be ignored. Meanwhile he was markedly reluctant to discuss exactly what his invention was. Asserting, probably quite rightly had the thing been all that he claimed, that to reveal details to all and sundry would be to invite theft, plagiarism and, consequently, loss to the United States as a whole, the amiable Giragossian could only repeat in glittering terms that his machine was all that he said it was, and probably more. Unlike the many fraudulent inventors before him, Giragossian was willing to allow the fullest possible examination of the project—by a team of top engineers and scientists appointed by the US government.

What really made things worse was that the general press cottoned on to the story and blew the whole project, known as the Garabed scheme, to the great American public. Giragossian freely admitted that his idea was not yet in a form which could be used to power America, but this was only a detail as, after its details were known, there would be no question of funds not being available from the taxpayers. Giragossian approached the Senate and Congress and the press supported his cause.

As 1918 opened, a Congressional committee made a preliminary investigation, passed a special Act of Congress, and the President appointed a team of five scientists to look into the machine and report on its merits. The chairman of the so-called Garabed Commission was James A. Moyer. The whole country was keyed up to a high state of expectancy as claims as to what the Garabed project could mean became more and more wild: free electricity, free water pumping, factory plant operated at a fraction of the usual cost and so on. Small wonder many people toasted the inventor.

In August the Garabed Commission completed its report. As those of wisdom had asserted from the beginning, the Garabed dream of free energy was based on the insecure foundation of observation by ignorance. Some large measure of its improbability had surrounded the whole matter when it was learned that Giragossian had declined to secure a patent for his idea as he felt the patent authorities would not give him protection. It was this that had inspired him to rush off to Congress and plead for direct help. In agreeing, of course, Congress by its unprecedented attentions actually fostered these suspicions and cast a slur upon regular legal means of protecting inventors through the services of the Patent Office. Now the Commission reported on what they found.

A flywheel!

Garabed's great machine consisted of a heavy flywheel which could be set in motion by means of a system of pulleys. The flywheel was mounted in bearings which were devised to keep friction to a minimum, and it was provided with a small electric motor driven by an equally small electric battery. The power generated by battery and motor was sufficient to overcome the friction of the machine and also the air resistance. Garabed claimed that the machine would start itself, but that it would take a very long time to run the flywheel up to its full speed. Because of this, the Commission reported, it was started for them by a strong man by pulleys and belts. After the machine had started, the battery was switched in and the machine would continue to revolve indefinitely with the flywheel developing a rim speed of around 100 feet a minute.

The power of the machine was measured by a device familiar to the so-called pony brake or power dynamometer which was loaded with weights until the machine was forced to stop. From this it was possible to calculate the horsepower. Apparently it took only one-twentieth of a horsepower from the motor to keep the flywheel running, but it took ten horsepower to stop it. From this, Giragossian concluded that he was actually producing energy. Today we would describe this as a constant speed flywheel.

Unable to differentiate between power and energy, the inventor could not realise that the energy was that put into it by the man and the motor over a period of time, and that all Giragossian was doing was stopping the thing suddenly and noting how great a power had been developed by expending in a few seconds the energy which had taken minutes to be stored up.

Giragossian demonstrated beyond the smallest doubt by his utter honesty at all times that he had set out to defraud no one. The opinion of the investigators was unanimous. Had the disillusioned man gone through the normal channels of applying for a patent, the Patent Office would soon have explained the fallacy of his reasoning and a great deal of time and public money would have been saved. No more was ever heard from him, and perhaps very wisely America has forgotten about Garabed more completely than about any other of its 'perpetual motion' frauds. Not that the sheer enthusiasm with which Giragossian pressed his

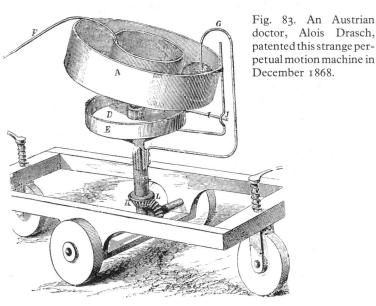

Fig. 83. An Austrian doctor, Alois Drasch, patented this strange perpetual motion machine in December 1868.

Fig. 84. From *The Mechanic's Magazine* published in London in 1829 comes this curious design for a 'self-moving railway carriage'. The inventor describes it as a machine which, 'were it possible to make its parts hold together unimpaired by rotation or the ravages of time, and to give it a path encircling the earth, would assuredly continue to roll along in one undeviating course until time shall be no more'. A series of inclined planes are envisaged in such a manner that a cone will ascend one (its sides forming an acute angle), and being raised to the summit, descend on the next (having parallel sides), at the foot of which it must rise on a third and fall on a fourth, and so on. The carriage, entered through the top, has broad, conical wheels and 'the most singular property is that its speed increases the more it is laden; and when checked on any part of the road, it will, when the cause of the stoppage is removed, proceed on its journey by mere power of gravity'.

claims upon the American public is a trait specifically reserved for that side of the Atlantic. Allow me to quote a letter sent to the British Patent Office and dated 15 September 1851:

'Gentlemen:
I pray you to take steps to make known that yesterday I completed my invention which will give motion to every country on the Earth;—to move Machinery!—the long sought in vain Perpetual Motion!!—I was supported at the time by the Queen and H.R.H. Prince Albert. If, Gentlemen, you can advise me how to proceed to claim the reward, if any is offered by the government, or how to secure the PATENT for the machine, or in any way assist me by any advice in this great work, I shall most graciously acknowledge your consideration.

'These are my convictions that my SEVERAL discoveries will be realised: and this great one can be at once acted upon: although at this moment it exists only in my mind, from my knowledge of fixed principles in nature: the Machine I have not made, as I completed the discovery YESTERDAY, Sunday!...'

From his opening sentence, the unsuspecting reader might suppose that the claimant had devised a new cathartic! So many thought that a great reward was theirs to claim; so very many were disillusioned.

12

Ever-Ringing Bells and Radium Perpetual Motion

In an earlier chapter, I tried to give some impression of the enormous effect that the discovery of electricity and electro-magnetism had on inventors during the last century. Static electricity became an object of fascination for all, and up and down the country and in the Mechanics Institutes which every worthwhile community provided, many were the young and old (the ladies as well!) who marvelled at demonstrations of generation, attraction, repulsion and sparks of countless sizes and colours. Electricity at this time was in its infancy and it was not until the late 1880s and the 1890s that the widespread use of it for lighting came into fashion—and with it the multitude of deaths and fires by electrical short-circuit. But long before that, Alexander Volta (1745–1827) discovered that two superimposed discs, one of copper and one of zinc, formed what he termed an *electro-motive couple* and that if you repeated this arrangement a number of times, alternating between copper and zinc discs, until you had a pile of discs, an electric current could be produced.

Volta's voltaic or column pile was the forerunner of the familiar dry battery of today. In subsequent years, numerous inventors experimented with the phenomenon which the pile displayed to discover the source of what we now call voltage. For a long while it was thought that the voltage developed in Volta's original wet battery was due to the contact between the copper and zinc discs, and the role of what is now termed *electrolyte* was not appreciated although many experiments were carried out, especially to determine whether the presence of wetness was essential or merely advantageous. In time it was to be determined that an *absolutely* dry pile would produce no voltage and that the degree of moisture was itself somewhat critical.

The first to construct a dry pile was an experimental scientist called Behrens[1] who used zinc and copper plates separated by discs of flint, the surfaces of which had been rubbed with the adjacent metal. Each disc was separated by gilt paper which had been soaked in a weak salt solution and then allowed to dry. Behrens made dry piles from 1803 onwards but it seems unlikely that he saw them as anything but intriguing experiments in the new sphere of electro-magnetism.

Next came the man who was the first to produce a working model as a philosophical demonstration of the dry pile. He was J. A. de Luc, a Fellow of the Royal Society, who had been working in London on voltaic cells as early as 1800. Between about 1806 and 1809, he constructed a number of dry piles.

De Luc's pile, as it came to be known, comprised a large number (1,200 to 1,600 alternations) of discs of thin sheet-zinc, paper and silver foil. Smooth writing or cartridge paper was overlaid with silver leaf and the sandwich cut into small round plates using a punch. Next, extremely thin sheets of zinc were covered with ordinary writing paper and the punching repeated. The pile was then

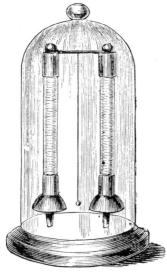

Fig. 85. De Luc's continuous chime.

[1]His work is described in the *Annalen der Physik*, vol. 23, 1, 1806 and also referred to in vol. 49, 47, 1815.

assembled in the order zinc, writing paper, cartridge paper, silver foil, so that the foil contacted the zinc and was separated from the next silver disc by two discs of paper. The pile was then clamped together very tightly on an insulated rod. The actual discs did not need to be more than $\frac{5}{8}$ in. in diameter.

De Luc arranged two such piles so that they were supported vertically side by side and insulated, one positive side down, the other negative side down. The tops of the two piles were connected together with a wire. By fitting a small bell at the bottom of each pile each making electrical contact with its end of the pile, and suspending a tiny brass ball between them by means of a slender thread of raw silk, de Luc found that the ball would alternate between the two bells, striking each in turn. The greater the number of discs, the more impressive the machine looked and the better it worked. De Luc called this his *perpetual chime* and ended up making such a chime with between 2,000 and 3,000 discs. Under normal atmospheric conditions, sufficient voltage is provided to keep the tiny ball traversing the short distance between the two bells. A pile of 2,000 pairs of plates operated continually for twelve years until it was stopped by an accident.

The first published account of his work appeared in *Nicholson's Journal* in 1810[1] and in this he stated that the Royal Society had refused to publish a paper of his in 1809 as a consequence of which his results had 'lost the merit of novelty'. Be this as it may, contemporary records show that de Luc had visited Behrens in Berlin and the latter's work undoubtedly antedated his.

Several copyists built perpetual chimes on de Luc's pattern. A Professor Silliman at Yale University reported[2] that he kept 'a set of these bells ... in Yale College laboratory for six or eight years unceasingly'. It appears that he decided that the point was proved at that stage and dismantled the device.

A professor in the Lyceum at Verona, Abbot Giuseppe Zamboni, was meanwhile engaged in his own versions of the dry pile.[3] His principal departure from the dry piles so far made was to introduce manganese dioxide which at that time was considered to be a substance of peculiar significance in that it gave up oxygen on

[1] *Nicholson's Journal*, vol. 26, June 1810, p. 113.
[2] John Phin, *The Seven Follies of Science*, p. 41.
[3] Described in Brugnatelli, *Giornale di Fisica ... del Regno Italico*, vol. 5, December 1812, pp. 424–46, and vol. 6, January 1813, pp. 31–43, and reported in Gilbert's *Annalen der Physik*, vol. 49, 1815, p. 35.

heating. This property no doubt played some part in influencing its selection by Zamboni who thereby anticipated its much later use as a depolariser in the modern dry battery. Giuseppe Zamboni's demonstration machine was more spectacular than that of de Luc (see Fig. 86), and his dry pile comprised discs of unsized paper coated on one side with zinc and on the other with a paste of manganese dioxide and honey. The manganese dioxide thus served not only as a depolariser but also to provide sufficient manganese ions for this element to become a voltaic partner with the zinc.

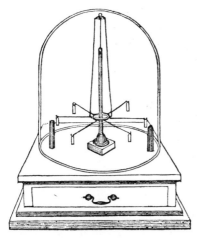

Fig. 86. Zamboni's perpetual motion.

Apart from his essentially visual demonstration model, the ever-practical Zamboni later published an account of the application of his form of dry pile to providing the power for a pendulum clock. Dr A. J. Croft of the Clarendon Laboratory in Oxford has discovered that at least one of these clocks is still in existence, but the pile has disintegrated. When last it was heard of a few years ago (written early in 1974), it was being restored.

One other man should be chronicled. This is George John Singer who published detailed instructions on how to make dry piles.[1] Singer's version of the de Luc pile used as a 'perpetual chime' comprised two glass insulators supporting two German silver bells each having a dry pile mounted on it. The tops of the piles were connected by a wire stirrup from which a striker hung on a fine thread.

[1] *Elements of Electricity and Electrochemistry*, London, 1814.

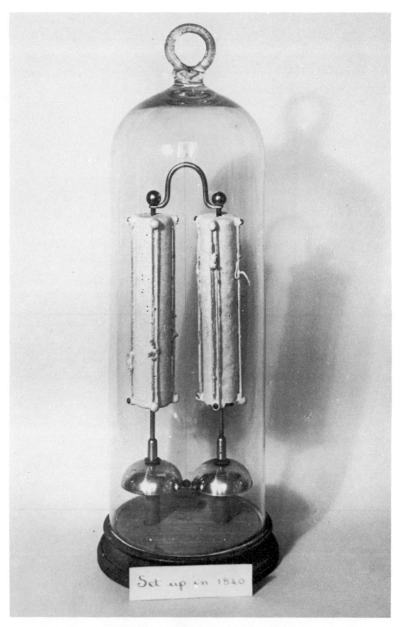

Fig. 87. The perpetual chime set up by Singer in 1840 and still in operation today. *Picture by courtesy of Dr A. J. Croft of the Clarendon Laboratory, University of Oxford.*

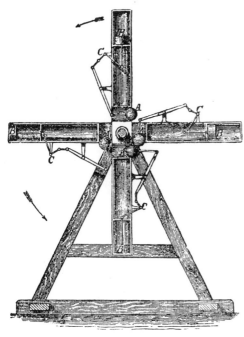

Fig. 88. George C. Philipps of Alleghany returned to out-of-balance weights for his perpetual motion machine in the 1860s. Heavy pistons in rotating cylinders were induced to slide back and forth by the weight of balls mounted on levers at the hub, motion being transmitted through levers and cranks. Friction and the equal moment of all weights about the centre regardless of position united to defeat the intended *perpetuum mobile*.

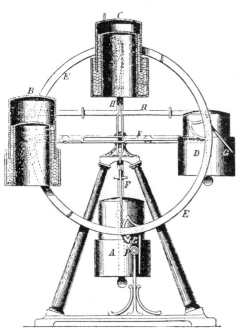

Fig. 89. A hydrostatic perpetual motion machine was patented in the 1860s. The weight of one double container (an inner air chamber floating in water in an outer vessel) was transmitted to the inner air containing chamber on its opposite side. Ingenious but ineffective.

There is at least one dry pile chime on the Singer principle still in existence—and still softly ringing its way across the years. This is preserved in the exhibition case of the Clarendon Laboratory of Oxford University under the care of Dr A. J. Croft who has traced its origins and made an assessment of its life. It was purchased from a London firm of instrument makers, Watkin & Hall, and a piece of paper bears the manuscript legend 'Set up in 1840'. The handwriting has been identified as that of Dr Robert Walker, the then Reader in Experimental Philosophy.

The Clarendon Laboratory's specimen is interesting in that the piles are compressed by string loops and the whole pile, loops as well, has been dipped in molten sulphur which acts as a seal and an electrical insulator. Already the piece has been in continuous motion for at least 134 years, the two bells being struck alternately at irregular intervals amounting to several times a second. The operation is only broken for occasional short periods when the humidity is high. It has proved impossible to determine exactly the voltage being produced but it is estimated to be in the order of one volt plus or minus 0·5, and the current at between one and 10 nono-amperes.

As regards the future life expectancy, this is difficult to estimate. Dr Croft believes that it can be measured in hundreds of years but there are already signs that the materials of which the striking mechanism is made are deteriorating. The striker, originally a spherical ball, is now quite noticeably worn with a 'waisting' having taken place just above its widest diameter corresponding to the point of contact with the bells. Beneath this, on the base of the mechanism, is a tiny but ominous pile of metallic dust.

At the most, then, the Clarendon 'perpetual chime' may prove to have a life of around 500 years. Within the terms of the accepted definition of the word *perpetual*, it is obviously a failure since not only is it wearing itself out, but it is dependent for its action on sources from within itself. Nevertheless, I feel that the creation of something which, provided it is kept reasonably safe from accidental or incidental damage, may continue to show movement for half a millennium, however it achieves that, is something deserving of more than passing mention.

The fundamental principle of the so-called de Luc dry pile was the construction of what was termed an *electroscope*. Various inventors, as I have shown, tried other varieties of exhibiting continual motion. One, Bohnenberger, achieved a similar but less spectacular result using a tiny strip of gold leaf between the two poles of the pile. Others concentrated on trying to better the construction of the pile itself, experimenting with such diverse materials as slices of walnut, beetroot and radish to form the discs. For the bemused reader today, the story loses a certain charm when I have to reveal that the edible voltaic pile was a failure.

For those who would wish to make for themselves a 'perpetual chime' that might well tinkle into a measure of eternity as a memorial to its erstwhile creator, practical instructions for making dry piles are to be found in the *English Mechanic* for 1915 and also in papers describing the use of dry piles as the power supply for an infra-red viewing device used during the 1939–45 war.[1]

The strange property of some substances to display the phenomenon of radio-activity was first discovered in Paris by the French physicist Antoine Henri Becquerel (1852–1908) in 1896 following a study of uranium. In 1903 Mme Sklodowska Curie discovered radium, and this material at once took over from electricity and electro-magnetism as the wonder discovery of the age. John William Strutt, the English physicist, built a fine foil electroscope which was gradually charged by the emission of radium rays until the point of discharge, whereupon the system worked over again.

The regular and precise charge and discharge cycle not unnaturally inspired some to use the principle for recording time, and a radium clock was described in the magazine *Work* for 25 June 1904:

'A radium clock, which is claimed to be able to time indefinitely, has been constructed. The principle is very simple, the registration of time being made in two-minute beats, while its function is to exhibit the dissipation of negatively charged alpha and beta rays by radium.[2] The clock comprises a small tube, in which is placed a minute quantity of radium supported in an exhausted glass vessel by a quartz rod. To the lower end of the tube, which is coloured violet by the action of the radium, an electroscope formed of two long leaves or strips of silver is attached. A charge of electricity in which there are no beta rays is transmitted through the activity of the radium into the leaves, and the latter thereby expand until

[1]*Electronic Engineering*, September and October 1948, T. H. Pratt and A. Elliott.
[2]The editor of *Work* obviously missed the fact that only beta rays are negatively charged: alpha rays are positively charged.

they touch the sides of the vessel, connected to earth by wires, which instantly conduct the electric charge, and the leaves fall together. This simple operation is repeated incessantly every two minutes until the radium is exhausted, which, in this instance, it is computed will occupy thirty thousand years.'   ·

A similar piece of equipment was exhibited at lectures given later in the same year by Sir William Ramsay, FRS. This was built to the designs of a Dr Hampson and was also the subject of a description in *Work*[1] as follows:

'By means of an exceedingly small quantity of radium salt, a feather is electrified. It bends away from the metal until it touches the side of the vessel, and looses its electrical charge, then springs back, and is again electrified, the process being repeated any number of times, practically like the swinging of a pendulum. Says the "Bazaar", a clock of this kind would be conceivably possible, and as it would persist as long as the radium retained its power, there might be a timepiece going for, say, 2,000 years, and never require to be wound up.'

On another device, the journal *Electricity* commented in its issue for 10 June 1904:

'In the radium clock of Harrison Martindale, an English physicist, we have to all accounts a solution of the perpetual problem of perpetual motion, which has so long provided our crank inventors with an object upon which to expend their surplus energy. In his apparatus a small quantity of radium, supported in an exhausted glass vessel by a quartz rod, is placed inside a small tube, to the lower extremity of which is attached an electroscope, consisting of two long silver strips. The activity of the radium causes an electric current to pass to the latter, which diverge until they touch the sides of the vessel, and are instantly discharged by earthed conductors. Falling together again, the strips receive a fresh charge, and the operation is repeated every two minutes. In theory, the action is calculated to continue until the radium is exhausted, a small matter of 30,000 years, so that the inventor is hardly likely to be called upon to recharge the device when it has run down.'

Apart from the ever-present hazard of radiation exposure, not

[1] 19 November 1904.

even suspected at that early period, the radium clock had a major disadvantage which, in its early form at any rate, rendered it somewhat short of its estimated performance. The difficulty lay in atmospheric ionisation which necessitated the placing of the device in a vacuum. In 1911, the principle was improved upon by one Herr Greinacher whose perpetual motion machine could work in open air and demonstrate to an audience its movements even when using a very weak radium compound. As little as three milligrams of radium bromide ($RaBr_2$) would keep the needle in motion for between five and nine minutes on one charge.

To the best of my knowledge, nobody ever set up a radium perpetual motion machine and left it to run; so, although the radio-active life of the material is considerable, no radium machine survives to substantiate the claims of its inventor.

The decay time of natural radio-activity is of such a prodigious length that scientists are today able to apply this to the dating of objects of archaeological and geological interest. The technique of carbon-dating makes practical use of the characteristics which the perpetual motionists could not.

13

Perpetual Motion Inventors Barred from the US Patent Office

A journal of outstanding historical value is the *Scientific American* which, under the editorship of Orson D. Munn, was to chronicle the inventive world and its activities from its Broadway, New York offices from 1845 onwards. Munn, who was also publisher, was an extremely thoughtful and intelligent man who commented freely on the affairs of the scientific world about him. A subject which occupied not inconsiderable space in the *Scientific American*'s pages over the years was perpetual motion. It was this journal which was shrewd enough to see through just about all of the devices put forward, and when the Garabed machine nearly brought the US Senate to ridicule in 1918 (see Chapter 11), it was the *Scientific American* which steered a sane course through the riptide of public feeling.

Munn saw that if we accepted as perpetual motion the harnessing of one of the variables of Nature, perhaps in the same way as Cox did in London, or tidal motors, solar generators or thermal motors, then perpetual motion was possible. But to conceive machines which would deprive an object of gravity whilst ascending, yet to restore it while descending, to cause masses of matter to act alternately in accordance with and in opposition to the laws of gravity, and to make water run up-hill—these, Munn decreed, were rubbish.

Nevertheless he acknowledged that the perpetual motion seekers had not altogether wasted their time, for while they had been learning that 'knowledge comes of experience' they had been led to consider more practical subjects and had in some cases made inventions which have proved truly beneficial to the world and profitable to themselves.

The position which the *ScA* held and, to a lesser extent, does still hold today, was by no means insignificant, and since just about anybody who had anything worthwhile to contribute to any of the sciences (and that included the follies of science) appeared to be brought to the notice of the journal, the editors usually managed to find editorial space for them. Editorial comment was provided where necessary, but more often than not the contributor was left to the varying mercies of his fellow readers—and they seldom failed to respond to a challenge.

As a result of an Act passed in England by the King in 1623, it became possible for the property and right of inventors in arts and manufactures to be secured by *letters patent*. The history of patents (from the Latin *pateo* meaning 'I lie open') goes back to the time of Edward III when patents are said to have been granted for titles of nobility. But with the patenting of inventions, the inventor might secure a monopoly of his idea for a period of time during which he could exploit it without fear of infringement by entrepreneurs. Thus it became the practice for every inventor to aim to secure his brainchild with letters patent. Many were the brilliant and far-reaching ideas of the early inventors. And then there were the perpetual motionists.

The earliest British patent for a perpetual motion machine was granted on 9 March 1635, and refers to 'skill of makeing engins, which being put in order, will cause and mainteyne their own mocions with continuance and without any borrowed force of man, horse, wind, river, or brooke, whereby many severall kinde of excellent rare worke may be pformed to the great good and benefitt of the comon wealth, the like cause and means of which continuance of mocion hath not been heretofore brought to pfeccion'. Unfortunately, at this early time of patent requirements, it was not necessary to furnish details or drawings of the device envisaged by the patentee, so we are left completely in the dark as to how this 'engin' was 'brought to pfeccion'.

The second patent, granted in 1662, is little better but does include a more definite statement to the effect that it is a perpetual motion machine.

'An engyne which, with the perpetuall mocion of itselfe, without the help or strength of any person or creature, will not only dreyne great levelle of vast quantities of water but also mynes of fifty fathom deep or more.'

Perhaps this was some sort of a syphon but, if so, to drain a fifty-fathom-deep 'myne' would have required a receptacle of somewhat greater depth.

During the ensuing years similar inventions occurred at odd intervals, until the middle of the last century when the Industrial Revolution brought with it a tide of theorists who believed that with machines varying from the ridiculously simple to the incredibly complex, perpetual motion was within their grasp. The majority of these fervent hopes for personal fame and fortune rarely distinguished themselves by even mechanical ingenuity, and occasionally the hapless inventor had simply bewildered himself by the complexity of his own device. So often the addition of another gear ratio, train of wheels or set of levers was believed to be the only thing required to overcome the friction already present.

The records of the British Patent Office show that as recently as the years 1901, 1902 and 1903, there were, respectively, thirteen, ten and nine applications for patents relating to perpetual motion. There have been some at intervals since. But from earliest times up to the end of 1903, over 600 patents were granted. Of these, only twenty-five pre-dated 1855, indicating the enormous impetus given to invention in general by the Victorian era.

Of a total of thirty-one applicants for patents during the years 1897 to 1900 inclusive, ten were English, eight American, three French, five German, two Australian and one each from Belgium, Russia and Austria.

The majority of these projects for obtaining perpetual motion relied on the force of gravity, the loss of equilibrium, specific gravity of floats and weights immersed in water or other liquid, ascension of receptacles inflated with air or gas under water, compression and subsequent expansion of gases, and of the surface tension of liquids.

It proves the undoubted faith and confidence in their brain-children, and perhaps a slight Frankenstein complex, that in many cases the inventors were careful to specify a braking system on their machines so that they might be stopped and prevented from a dangerous increase in speed!

It was in 1880 that the *English Mechanic* related how an inventor from Yorkshire had applied for a patent, but had stopped short at the preliminary stage, no doubt exasperated at the obdurate behaviour of his model. He did, though, leave us a description:

'The engine is perfectly and entirely self-acting, creating its own power, without any agency, except the pressure of the atmosphere, which the construction and arrangement of the engine converts into a substantial active power which is limited only to the strength of the material, and supplies a continuous source of power by its action. The utility of the self-acting atmospheric engine will effect a saving of considerably over £120,000,000 per annum in the United Kingdom of Great Britain and Ireland, in fuel, labour and materials.'

The quest for the end of the rainbow crossed the Atlantic and the United States Patent Office in Washington became the receiving ground for American attempts at the unattainable. When this building burned down in 1836 it was estimated that about ten patents for perpetual motion machines had been granted and were on file.

As early as 1828, the journal of the Franklin Institute in Philadelphia published a lengthy explanation of why perpetual motion could not be perpetual. When scientists came to accept the principles of the conservation of energy as set out in Chapter 2 above, the belief in the possibility of there being perpetual motion began to recede. Even so the US Patent Office was slow to act, and it was not until well into the second half of the last century that the applicant for a patent regarding a perpetual motion machine was required to submit a working model. Almost a hundred years earlier, in 1775, the Parisian Academy of Sciences refused to accept schemes for perpetual motion.

Clifford B. Hicks, who is editor of *Popular Mechanics Magazine*, has collected together a random assortment of reckless and laudatory newspaper items on these inventors and their claims.[1] The *Gazette* published in Philadelphia in 1829 contained a glowing editorial as follows:

'We were much gratified yesterday with the result of an examination of a self-moving machine, which may be seen at Bowlsby's

[1] *American Heritage*, vol. 12, April 1961, pp. 78–85.

Merchants' Hotel, in Slater Street, and which the inventor calls perpetual motion. We have no doubt of it being nearer a perpetual self-moving principle than any invention which has preceded it, and as near as any we shall ever see. The great merit, aside from its practical uses, is its simplicity, and the certainty and readiness with which you perceive that it covers no trick or deception.'

In 1854, the *New York Journal of Commerce* waxed enthusiastic about a machine invented by a gentleman named J. G. Hendrickson. Displaying a simple ebullience, the newspaper wrote:

'The model was in our office yesterday, and attached to some clockwork, which it turned without once stopping to breathe. We see no reason why it should not go until worn out! After a careful examination, we can safely say, in all seriousness, that the propelling power is self-contained and self-adjusting, and gives sufficient force to carry ordinary clockwork, and all without any winding up or replenishing.'

After so tantalising a write-up, one might justifiably expect there to be a follow-up. It came—in 1868:

'About fourteen years ago we published the first description of a machine invented by Mr. James G. Hendrickson of Freehold, N.J. ... [and] we saw no reason why it would not go until it was worn out. The inventor was an old man, who had spent his whole life in pursuit of the object he had now attained. He was invited to be present at various fairs and exhibitions of new inventions, and wherever he went, his machine formed one of the chief attractions. The professors were all against him. Accordingly, Mr. H. was seized at Keyport, N.J., for practising "jugglery" under the "Act for Suppressing Vice and Immorality". To expose the supposed trick, an axe was brought, and the cylinder splintered into fragments. Alas! There was no concealed spring, and the machine had "gone of itself". He made a new machine. His model once more completed, was constructed of brass, hollow throughout. The moment the blocks were taken out, the wheels started off "like a thing of life"; and, during ten months, it never once stopped. The inventor had perfected two new machines, and made a very comfortable livelihood exhibiting them, prosecuting his efforts meanwhile to secure his patent. Age crept upon him, however, before this point was reached; and last Saturday afternoon he breathed his last at Freehold. The night after his death his shop was broken open, and both models stolen.'

It is a sad story, but the end conveniently disposes of any possibility of learning the secret of the mysterious mechanism. Orffyreus destroyed his wheel, and the remarkable aircraft engine with which I had dealings only a decade ago (the story is related in Chapter 17) had to be dismantled 'to stop it running'. The evidence, I hope you have been able to detect, is always missing! Another case of the vanishing machine resulted from the startling article published in *McClure's Magazine* in 1899. In a discourse written by Ray Stannard Baker, later to become an eminent author, historian, government agent and a friend and confidant of Woodrow Wilson, information was given regarding a machine invented by Tripler of New York.[1] If the machine had worked, it would have made perpetual motion not only possible but virtually inevitable. Baker, you will recall, arranged for two scientists to visit Tripler's laboratory. The time proved inconvenient for Tripler and so the scientists never did get around to examining the machine. Clifford Hicks makes a valid comment on this. 'There never was,' he says, 'indeed there never is, a convenient examination for such devices. This is almost another law of physics.'

In 1902 a civil engineer presented a paper at a meeting of the American Association for the Advancement of Science. In it he implied the possibility of perpetual motion in the following terms:

'The Second Law of thermodynamics is fallacious. The *effect* of an operation *can always be reversed*, and when produced by an operation which is made irreversible by the unrestrained or unbalanced action of some particular element or elements, can be reversed by another irreversible operation made irreversible by the unrestrained action of another element or elements having an opposing action to the first mentioned element or elements. I wish to contribute the above statement to Physical Science.'

It fell to a mathematician named Mudie, writing in his book *Popular Mathematics* published in 1836, to become the first prognostic of extra-terrestrial man-made perpetual motion. Although his sums were incorrect and the term 'escape velocity' had yet to

[1]Tripler's system of the liquefaction of air is described in Chapter 10 above.

be coined, his notion took about 120 years to come to a measure of reality:

'It is not difficult to calculate (upon mathematical principles) that if we could give any piece of matter a motion round the earth at the rate of about five miles a second, or 1,800 miles an hour, and keep up the motion at this rate, we should overcome the gravitation of that piece of matter. This is what may be regarded as the possible case of perpetual motion.'

But perpetual motion applicants continued and their wild notions served the interests of less hare-brained inventors in a novel manner by the fees which they so willingly paid to the US Patent Office. A fool and his money, it is often said, are easily parted. The story is related by the *National Car & Locomotive Builder* for 1891:

'Some of the most ingenious and persistent men are laboring on the hopeless task of devising perpetual motion appliances. Our educational system is in many respects responsible for so much mental energy being wasted upon fallacies. If natural philosophy and elementary mechanics received the attention in common schools that their importance demands, there would be fewer persons pestering their friends to supply funds for the development of apparatus intended to cheat nature's laws. Ignorance of the laws of nature is, no doubt, responsible for the majority of perpetual motion devotees, yet some men who are well educated become pursuers of the chimera. It is frequently difficult to understand the mechanical fallacies that creep over what are otherwise bright intellects. Electricity seems to be deceiving many men and leading them into the belief that by means of this mysterious force more power can be received than what is given. Since the electric lighting and electric transmission of power era began, there has been a great increase in the applications to the Patent Office for protection of what are electrical perpetual motion machines. For years the Patent Office income was considerably augmented annually by the receipt of fees from inventors of perpetual motion machines, but no fees are now accepted from men working on that kind of apparatus. A printed circular is sent to applicants saying that nothing short of a working model would be received, and as the inventor never gets a model of this kind to work, he can do no more with the Patent Office.'

Meanwhile, the perpetual motionists came, and went. One would think that after so many attempts, the very improbability of it would have deterred even the most simple-minded inventor. It certainly should have deterred men who might reasonably be expected to be possessed of more than average sagacity. A correspondent of the St Louis *Globe-Democrat* related a story which was reprinted by the *Scientific American* in its issue for 23 May 1891:

'A few months ago a New York lawyer went to Washington with parts of a machine, and had quite a controversy with the [patent] office because the patent was refused. He insisted that he had seen the machine in operation, that it was running day after day, and keeping a cider press going to boot. There was no deviating from the rule. The lawyer went back to New York, saying that he would produce the machine. He was not seen again until the centennial celebration lately, when he reminded the examiner of the case and told him how he had been fooled. At the time of making application the lawyer really believed that his client had discovered the long-sought principle. But when he got back to New York and told him that the patent had been refused, the client confessed. The perpetual motion was no motion at all. Power was concealed in the cider press. It ran the press, and the press made the perpetual motion machine go too. The inventor had been charging 10 cents admission to see perpetual motion. He had fooled the public and his lawyer, and he hoped to slip through a claim.'

The American Patent Office's requirement for a model, intended to deter misguided inventors, did not altogether have the desired effect, and many a patent attorney, having tried hard to convince his client of his waywardness, would find himself compelled to prepare specifications and claims relating to a machine which could never be. An inventor who showed such commendable imperviousness to reason (if one is going to be ignorant, then one may as well make a thorough and studied job of it) would not be dismayed at the thought of having to spend hundreds and perhaps thousands of presumably hard-earned dollars to try to build an operative perpetual motion machine model.

The peculiar circumstances which concerned this particular type of invention became of great concern to the Patent Office in Washington. The procedure of accepting an application and then allowing the inventor up to a year in which to prove his idea with a working model was unwieldy since, of course, none of the applicants ever returned and the mechanism of the patent system was becoming cluttered with unconsummated applications. In 1911 the Commissioner decided that, for perpetual motionists, a new ruling would come into force which demanded a working model before even an application could be completed. Although this did not prevent misguided inventors from spending their money (and quite frequently other people's) on trying to make models, it at least saved them the government fees on a hopeless application.

And so each potential solver of the impossible, when he approached the Patent Office, was given a printed circular. It read:

'The views of the Patent Office are in accord with those of the scientists who have investigated this subject and are to the effect that such devices are physical impossibilities. The position of the Office can be rebutted only by the exhibition of a working model. Were the application to be forwarded to the Examiner for consideration, he would make no examination as to the merits, but his first action would be the requirement that a working model be filed.

'In view of all the circumstances, the Commissioner has instructed that applications for patent on Perpetual Motion, complete in all other particulars, shall be held in the Application Rooms as incomplete until a working model has been filed. Such model must be filed within one year from the date of application, or the application will become abandoned. The Office hesitates to accept the filing fees from applicants who believe they have discovered Perpetual Motion, and deems it only fair to give such applicants a word of warning that the fees paid cannot be recovered after the case has been considered by the Examiner. For these reasons it has been thought best to meet the inventor at the threshold of the Office, and give him an opportunity to recover the moneys paid into the Office, in the event of his failure to comply with the requirement.'

Even this did not prevent the perpetual motionists from pursuing their notions. All it served to do was to deprive them of the protection of a patent. At this point in time it can be argued that this was neither a drawback nor a deterrent. Bright-eyed men imbued with success still fast-talked their way into money through investors and indulged in personal or communal deceit.

Whether it is a sign of the times in which we live wherein other calls upon our leisure time have effectively deprived man of both the time and the wherewithal to think, or whether at last the mechanical perpetual motionists are a dying breed, those who quest for the great automatic power are very few and far between today. During the investigation for the BBC television programme which was the root cause of this book, only a handful of people replied to advertisements for perpetual motionists to come forward and be counted. Of these only two claimed to have ideas and only one was hard at work trying to disprove Stevinus and his sixteenth-century discovery of equilibrium on inclined planes.

14

Rolling Ball Clocks

Clocks which worked by rolling balls should not be considered as in any way perpetual motion machines. The descent of the ball, particularly in the Congreve pattern of clock, serves as a pendulum escapement to regulate the letting down of spring power. The length of time taken by the ball to pass from its highest point to its lowest is a constant which is equal to a number of beats of a pendulum. However, rolling ball mechanisms, whether driving timepieces or not, were sometimes conceived as perpetual motion machines.

John Evelyn, the diarist, leaves us his description of such a mechanism which he saw on 24 February 1655:

'I was shew'd a table clock whose ballance was onely a chrystall ball sliding on parallel wyers without being at all fixed, but rolling from stage to stage till falling on a spring conceal'd from sight, it was throwne up to the upmost channel againe, made with an imperceptible declivity, in this continual vicissitude of motion prettily entertaining the eye every halfe minute, and the next halfe giving progress to the hand that shew'd the houre, and giving notice by a small bell, so as in 120 halfe minutes, or periods of the bullet's falling on the ejaculatorie spring, the clock part struck. This very extraordinary piece (richly adorn'd) had been presented by some German Prince to our late King, and was now in possession of the Usurper,[1] valu'd at 200*l*.'

Samuel Pepys also recorded a similar piece—indeed, it may have been the self-same item—in his diary for 28 July 1660:

[1] Oliver Cromwell.

'To Westminster, and there met Mr. Henson, who had formerly had the brave clock that went with bullets (which is now taken away from him by the King, it being his goods).'

In the years to follow, it was by no means uncommon to find a clock beating time by the descent of a ball or 'bullet' along an inclined plane every half minute. Wood, in his *Curiosities of Clocks & Watches*, says:

'Gainsborough, the painter, had a brother who was a dissenting minister at Henley-on-Thames, and who possessed a strong genius for mechanics. He invented a clock of very peculiar construction, which after his death was deposited in the British Museum. It told the hour by a little bell, and was kept in motion by a leaden bullet, which dropped from a spiral reservoir at the top of the clock into a little ivory bucket. This was contrived so as to discharge it at the bottom, and by means of a counter-weight it was carried up to the top of the clock, where it received another bullet, which in turn was discharged like the former.'

If we accept this description for what it is, then our dissenting minister had captured that will-o'-the-wisp, *perpetuum mobile*. But the clock was really driven by a descending weight whose energy was bridled by the rolling ball escapement.

Rolling ball clocks were referred to several times in works on horology which appeared during the seventeenth and eighteenth centuries. The first description of such a device is contained in Caspar Schott's work *Technica Curiosa* published in Nürnberg in 1664. Schott studied with Athanasius Kircher in Rome but spent much of his life working in Augsburg. Unfortunately, Schott's descriptions of his many mechanisms defy adequate comprehension. However, Book IV of his *Mirabilium Mechanicorum* contains an illustrated description of a perpetual motion machine driving a clock which was invented by Wilhelm Schroeter, a sixteenth-century priest. In this unusual mechanism, six balls are contained in a channel on top of the casework. They fall, one at a time, through an opening on to a wheel mounted below which has cup-like receptacles around its periphery. Guards are provided to prevent the balls falling out of the containers as the wheel rotates until each is allowed to fall out into a passageway to communicate to a second wheel. This in turn feeds each ball into a third wheel. Now each of the three wheel-carrying shafts carries a second bucket-wheel, only these three wheels are of larger diameter than the first set. From the bottom of the lowest wheel on to which each ball falls, it runs along a guide to a bucket at the bottom of the lowest wheel in the second set of three. This systematically lifts each ball back up, passing it from one wheel to the next above it, until the balls emerge back in the recess in the top of the case. Schott believed that the continual motion of the balls down and then up again would operate a clock train comprising two wheels and two pinions driven by extensions to the self-same three axles or shafts. Strange to say, the very geometric logic of the system would have made it quite impossible for even one ball to find its way back up to the top even with the maximum of five available balls on the driving (descending) side, since the designer had arranged that the return wheels were of larger diameter than the driving ones. The reason for this was to allow space for the balls to drop on the drive side, but at the same time to allow the transfer of the ball from wheel to wheel on the return side. This meant that each ball on the return side was at a greater distance from the axle of the wheel than a similar ball on the descending side. By the suggested laws of self-turning wheels, as described in Chapter 4, this was quite impossible.

Schott provided ideas for at least two more perpetual motion machines. One is a winding device with an endless belt powered by balls. These are deposited on the descending side of the belt which carries little buckets. At the bottom, the balls are simply lifted to the top to do their work again.

The next to give some consideration to making clocks which moved by the use of rolling balls was Francesco Eschinardi, an Italian Jesuit and mathematician who was born in 1623 and died around 1700. In a work entitled *Ragavgli del Padre Francesco Eschinardi*, the full title of which translates as 'Account of the Father Francesco Eschinardi ... Given to a Friend in Paris', is a description of a time-indicator intended for use in navigation. The author states that early in 1670 he had discussed a clock which he had designed for this purpose with the Prince Leopold de Medici. He scorned weight-driven clockwork as then used in clocks because he considered it an inaccurate method of driving a time-recording instrument. Here he was displaying a basic awareness of the problems of time-keeping at sea—problems not solved until the invention of Harrison's marine chronometer in

1735. In place of springs or heavy weights he used one arbor carrying a crown wheel and a drum around which hung a chain of small
cup-like buckets. Through an aperture above the chain, a ball
dropped into one of the buckets and travelled downwards, turning
the shaft as it went. At the bottom, the ball dropped out of the
receptacle and landed on a spring-loaded trigger which allowed
another ball to drop on to the chain. As the second ball descended,
it wound up the other ball so that when the second ball dropped
off, the first ball was freed to drop back on to the belt, winding
the second ball—and so on. Motion that was believed to be perpetual ensued. It is unclear exactly how Eschinardi intended his
clock to function but it appears that when each ball fell off the
bucket chain and landed on the spring-loaded trigger, it wound
up the time-keeping part which was presumably spring-powered.
One cannot but admire the assiduous self-delusion of the
respected Father who expected his descending ball not only to
have power enough to spare after turning the shaft to raise up a
similar ball, but to be able to fall the final distance with sufficient
impetus to keep a spring wound.

Grollier de Serviere (1596–1689) described no fewer than four
rolling ball clocks in his *Recueil d'Ouvrages Curieux de Mathema-
tique et Mechanique* which was published in 1719. It was largely
through this authoritative volume that this type of clock came to
be known to horologists.

M. Grollier's first clock, temple-like, comprises a base containing the clock, and a dome supported on six columns, the whole
hexagonal in plan form. Around the outside of the columns wind
two parallel spiral tracks of brass along which a metal ball slowly
rolls from top to bottom. By some method not indicated, the
motion of the ball along its declivous path imparts motion to the
clock. His second clock is similar, except that the ball is visibly
returned to the top of the slope by a bucket-like system which
replaces the ball into a trough at the top, then collects it at the
bottom again, and so on.

The third clock again appears to rely on the power provided
by the descent of rolling balls, only this one has a single-plane
or zig-zag track and uses two brass balls. When one ball has completed its journey, the second starts its way down. At the bottom,
it frees the first ball which, in the interim, has been conveyed up
to the top once more.

Fig. 90. Clock invented by Grollier de Serviere in Lyons and apparently operated by a rolling ball perpetual motion system.

Grollier's fourth clock reverts to temple-shape, only this time the style is square. Once more the clock is in the base, this time with two faces, one each for the hours and the minutes. The dome is shown supported by six columns, four along the front side and one at each of the rear corners. Around the front columns runs the 'scenic railway' for the balls. Once the first of the two balls reaches the bottom, it enters the base of an Archimedean screw which revolves and carries the ball up to the top again.

All the clocks so far described exist only in early works on horology and there is no evidence to suggest that they were ever built. They owed their existence to the fertile minds of generally wise men.

Until its destruction in 1943 as a result of an air-raid, the Hessisches Landesmuseum in Kassel, Germany, had on display a rolling ball clock known as the *Perpetuum Mobile.*

The clock was made by Giuseppe Campani, the famed inventor and clockmaker of Rome who also made lenses for telescopes. He lived and worked in Rome from about 1650 until his death in 1715. Campani is renowned for the invention of several unusual clocks, and the Perpetuum Mobile was purchased from him by the Landgrave Karl of Hesse during a visit to Italy between December 1699 and April 1700. It was described fully by Schminke[1] as follows:

'The Campani clock is operated by means of two alternating balls which run over brass inclined grooves. As soon as one ball has completed its run, the other ball is released by the double-spoon. When the run is completed, the spoon drops the first ball into the groove, and this completes another run, and so on. Each ball completes its run and is ready for another in thirty seconds, and it is by means of this motion that the dial, placed at the top of the clock, is moved forward.'

This interesting piece stood 1·80 m (42·5 in.) high, about 0·75 m (29·5 in.) wide and 0·40 m (15·5 in.) in depth. It was contained within an ebony cabinet having glazed sides and simple ormolu decoration, including gilt bronze capitals and bases of four columns, a gilt bronze balustrade and gilt bronze draped wreaths on the front and sides of the supporting cabinet.

Upon cursory outward examination, the clock was apparently kept in motion by a brass ball which descended continuously along a helix of eight curved and silvered rails coiled within the frame of the case. Only one ball was visible at any time and the ball tracks were backed by mirrors.

In truth, this was a cleverly engineered fraud. The thing was operated by three springs contained in clock-type spring barrels,

---

[1]Friedrich Christoph Schminke, *Beschreibung der Hochfurstlich Hessischen Residenz- und Haupstadt Cassel*, Cassel, 1767, p. 165.

these being enclosed in a concealed space behind the mirrors. An essential feature of this clever clockwork mechanism is the absolutely silent escapement which Campani also used to provide the completely silent operation of the crank-lever used in an earlier silent night-clock. This escapement comprises a bar balance pivoted centrally and arranged at the top of the plates of the clock-work. Each arm of this balance is half the distance between the top and bottom of the helical path, or about as high as the front plate of the movement, and terminates in a spoon-like receptacle with the opening facing to the right. This balance rotates continu-ously in an anti-clockwise direction. Driven with this balance by pinions is another, smaller balance comprising two rods pivotally mounted at their centres and carrying two sliding weights. The speed of rotation of this balance and, thereby, the large bar balance, can be regulated by adjusting the position of these weights: further out for slower, closer in for faster.

The Perpetuum Mobile was set in motion by placing one of two balls into the upper bar balance spoon and placing the second ball in a trough provided at the start of the channel into the helical path. The ball ran along the groove to the bottom whereupon it fell on to a detent and came to rest in a position whereby it could be collected by the passage of the bar balance's receptacle. The time taken for the ball to descend is exactly thirty seconds which is the time taken for the bar balance to make one revolution, so that as the ball disappears at the bottom of its path, it seems to reappear back at the top. The clock hands receive an impulse twice a minute from the balance mechanism.

The success of this clock apparently lay in the subtleness with which it was executed. Well-hidden, silent clockwork combined with the minute hand which advanced in thirty-second jumps in time with the ball movement heightened the deception. There is, though, no indication that its inventor actually named it the Perpe-tuum Mobile, and this may have been a name given to it at a later date. At a time when novelty clocks were in high esteem, Giuseppe Campani probably schemed up this mechanism purely as witness to his ingenuity at fabricating mystery clocks. We may well do him an injustice by assuming him to have courted deceit.

Nobody ever suggested that the famous Congreve clock with its rolling ball and tilting platform was anything to do with per-petual motion, nor even an accurate timekeeper, but people are content to watch it, marvel at its operation and endow it with mystical powers. Sufficient to say that hand-made reproductions of it were produced in 1973 and found a ready market at a price in excess of £1,500.

15

Perpetual Lamps

This is a very abstract, if not altogether abstruse by-way of perpetual motion and justifies inclusion here on the grounds that the process of combustion is an energy-consuming form of power and the perpetual lamp, if it really existed, must have had the ability to create more combustible material to add to its substance during its very combustion, unless it was fuelled from a prodigious source.

Light has been a subject for reverence since time immemorial. In the form of the sun, its worship is as old as Man. To light which could be man-made at will and be used to banish the devils to the shadows was ascribed greatness and sublimity in ceremonials of both ritualistic and religious character. The lamp has carried down over the centuries much mystique, myth and magic, and it is therefore not surprising to find that it played an important part in the sepulchral rites of the ancients who placed lighted lamps in the tombs or vaults in which the dead were laid. These lamps were carefully tended and kept continually burning. A legacy of this survives with us today with the ever-burning flame on the tomb of the Unknown Warrior, a practice and connection which is to be found in many parts of the civilised world.

For references to lamps and to light, we have only to turn to the Bible to find them in abundance. Exodus, Chapter 27, details what the Israelites were required to do to make the tabernacle, and verse 20 reads:

'And thou shalt command the children of Israel, that they bring thee pure oil olive beaten for the light, to cause the lamp to burn always.'

The Third Book of Moses, Leviticus, contains an almost word for word repetition of this (Chapter 24, verse 2). It may be that these lamps were to be filled periodically: there is certainly no reference to their burning *perpetually* without extra fuel.

Some authors have claimed, however, that in ancient times lamps were constructed which did burn perpetually and needed no attention. Addison in number 379 of *The Spectator* related an anecdote based on this legend. He told how someone had opened the sepulchre of Rosicrucius.[1] There he discovered a lamp burning which was immediately struck to pieces by a clockwork statue. Addison added that the disciples of this visionary claimed that he had made use of this artifice to show that he had re-invented the ever-burning lamps of the ancients. This same story, without the attribution to Rosicrucius, is told in detail by Bishop Wilkins[2] who devotes some considerable space to a discourse on these lamps, their probability, and their likely construction.

The Romans are said to have preserved lamps in some of their sepulchres for centuries, and there are numerous legends of their never-dying combustion. During the papacy of Paul III (1534–40), Pancyrollus mentions that one was found in the tomb of Tullia, daughter of Cicero, which had been closed up for 1,550 years, but which was 'permanently extinguished upon the admission of new air'.[3] At the dissolution of the monasteries, a lamp was found which was said to have been burning for 1,200 years. Two lamps of this supposed type are preserved in Leyden Museum in South Holland. Wilkins tells of Cedrenus who in Justinian's time (AD 527) heard of a lamp found in an old wall at Edessa, Syria (later renamed Justinopolis) which had remained there for more than 500 years. Since there was a crucifix placed by this one, it was assumed that ever-burning lamps were also used by Christians. Saint Austin mentions one of these lamps in a temple dedicated to Venus which was always exposed to the open weather and could never be consumed or extinguished.[4]

[1] Rosicrucius, or Rozenkreutz, who is alleged to have lived in the second half of the fifteenth century and formed a secret society of mystics and alchemists called the Rosicrucians.

[2] *Mathematical Magick* by I. W., M.A. (John Wilkins, Bishop of Chester), London, 1648, section 'Subterraneous lamps, divers historicall relations concerning their duration for many hundred years together'.

[3] *Ibid.*

[4] *Ibid.*

Bishop Wilkins relates several other instances as follows:

'But more especially remarkable, is that relation celebrated by so many Authours, concerning [Maximus] Olybius his lamp, which had continued burning for 1500 years. The story is this. As a rustick was digging in the ground by Padua, he found an Urne or earthern pot in which there was another urne, and in this lesser, a lamp clearly burning; on each side of it [the outer urn], there were two other Vessels, each of them full of a pure liquor, the one of gold, the other of silver ...

'Baptista Porta tells of another lamp burning in an old sepulchre belonging to some of the ancient Romans, inclosed in a glasse viall, found in his time, about the year 1500, in the Isle Nesis, which had been buried there before our Saviours coming ...

'In the Tombe of Pallas [son of King Evander] who was slain by Turnus in the Trojan war [1184 BC], there was found another burning lamp, in the year of our Lord 1401. Whence it should seem that it had continued there for above 2600 years; and being taken out, it did remain burning, notwithstanding either wind or water, with which some did strive to quench, nor could it be extinguished till they had spilt the liquor that was in it.'

The Italian physician and philosopher Fortunio Liceti (1577–1657) made a life-long study of the apparent phenomenon of long-burning lamps and wrote a book in which he collected together a large number of stories concerning lamps said to have been found burning in tombs and vaults.[1] In more recent times, Antoine Frederic Ozanam, the French scholar who lived between 1813 and 1853, published an extended discussion on the subject.

Perpetual lamps, as a concept and irrespective of whether or not they truly existed, represented a peculiarly innocent philosophical notion of delighting future ages and forever keeping away the evil spirits which lurked in the darkness. Mark Twain tells of a sacred flame that had been burning for so many centuries that its origins were forgotten.

Many attempts have been made to explain the supposed facts behind the claim that the ancients were able to construct perpetual lamps. The light sometimes seen on the opening of ancient tombs, suggests Bishop Wilkins, may have been due to the phosphore-

[1]Fortunius Licetus, *De Lunae Subobscura Luce Prope Conjunctiones*, Utini, 1642.

scence which is a well-known by-product of the decomposition of animal and vegetable matter. Decaying wood, under certain conditions, and dead fish are familiar objects which give out a light that is sufficient to render dimly visible the outlines of surrounding objects. Were such a light to be seen in the vicinity of an old lamp, Wilkins postulated, the impression might arise that the lamp had actually been burning, the draught on opening the tomb having blown out the flame. He makes no mention of the fact that aside from the availability of fuel for the flame, one would expect there to be an air supply to have supported combustion, and so his explanation, strange as it may sound, might just hold a little water.

Phosphorescent substances such as barium sulphate or Bolognian phosphorus, as it was once called, lead some people to assume that the 'art' of making perpetual lamps had been re-invented, but of course this substance loses its phosphorescent qualities after being kept in the dark for some time and periodic exposure to bright sunlight is necessary to re-activate its luminosity.

Although we can examine phosphorus (discovered in 1669 by Brandt in Hamburg) and various luminous materials as possible sources of explanation for the ever-burning lamp, all are quite incapable of the feats attributed to the lamp which burns for such a prodigious time as the legends would have us believe.

The perpetual lamp is in chemistry the counterpart of perpetual motion in mechanics. Both violate the fundamental principle of the conservation of energy. The ever-burning lamp is thus quite as absurd as a self-moving wheel or a waterwheel activating itself by feeding its own mill race. In the absence of a technology which could synthesise the effect of such a feat, either by the refraction of light by lenses or fibre-optics, or by radio-active degradation, one must conclude that reports of perpetual lamps were probably based upon mere errors of observation. We might wish that tomb-openers in the past had generally been a little more diligent in their task of recording just what they saw. Having made that comment, the only faintly jarring note is that Liceti was able to find so many apparent instances of lighted lamps in long-closed tombs.

So what can one make of these claims to have found burning lamps? Can there somehow, anyhow, be any truth in the reports? First, then, we can discount the probability of a lamp being filled with a 'liquor' which would enable it to burn for a very long time (in ancient times, a temple lamp was made which would burn for exactly a year, requiring refilling with oil just once at the prescribed time). Next comes the question of observation. To superstitious people, the presence of a lamp in a tomb, itself by no means uncommon or unlikely from our knowledge of the funeral rites of times past, might have implanted the belief that light was actually being emitted by the lamp. There remains the consideration of two points: the undoubted antiquity of the lamp as an artifact, and the association of these enigmatic lamps with the dead.

So far, we can establish that ever-burning lamps were sought after as a symbol or artifact of life itself since very distant times, that their manufacture with a fixed volume of fuel was impossible, and that their existence was therefore improbable. However, there is only one possibility and this results from the reference in Wilkins' work to the discovery of such a lamp recorded by Baptista Porta as having been found on the Isle of Nesis.

Nesis, later called Nesita, is an island on the coast of Campania in the bay of Naples which, according to the historians, has always been famed for its asparagus. The island was visited and written about by at least two capable observers. One was Lucan who was put to death in AD 65 when in his twenty-sixth year, and the other was Statius who died about AD 100. Both spoke of the island's air as 'unwholesome and dangerous'. Was this sulphurous gas of volcanic origin, or was it due to the natural expiration of methane gas? If so, then a fissure of natural gas might have been terminated with a lamp body and ignited. Again, if the lamp had been of sufficient size, water could have condensed out of the gas and appeared as a 'pure liquor' through which the gas would bubble to ignite at the mouth of the lamp.

Such a lamp would serve as a burning nozzle for a considerable supply of gas, but in order for it to burn a supply of fresh air would have been necessary. There would have been no trouble with the example mentioned by Saint Austin in the Venus temple quoted by Wilkins, but the allegedly subterraneous lamps pose another problem. There is no reason why a supply of fresh air should not have been piped into a tomb, and the foul air from combustion vented by a similar pipe, and such pipes might pass undetected by the efforts of the unsophisticated tomb-plunderers—and most of the tomb-discoverers in these early days were no archaeologists but were peasants bent on securing what they could for their own benefit.

One only has to consider the enormous resources of natural gas in the North Sea, for example, to conjecture how long one small burner might be fed from the reservoir. A million years? Probably more. A small borehole or natural fissure, closed over with a block of porous material such as pumicestone, both to reduce pressure and provide a burner, and contained within a lamp body might burn through storm and tempest for the rest of meaningful time. Should it be possible to accept this as a likely solution to actual corroborated cases of perpetual lamps being found in tombs (and these have yet to be established), one would expect the fount of natural gas to have been discovered first, then deified with its restyling as a mausoleum.

If one assumes that such lamps did exist (and I emphasise that the authoritative evidence of their discovery is lacking), then a logical question arises as to why it is that no further lamps are on record as having been found since the sixteenth century? We might explain this by first of all asserting that the sepulchral lamp would have been the exception rather than the rule, and that assuming it to have required a supply of fresh air, it would have lain near to the surface and would probably have been marked with an obvious exterior in the form of a monument. It is then convenient to assume that since the majority of recorded instances were located in areas which were well known to the plunderers, they were all quickly exploited in the Middle Ages.

If one denies their existence, then it can be attributed to the wisdom of later years and the more orderly procedures of archaeology. Perhaps in some subterranean cavern so far undiscovered there burns a lamp yet, waiting to shed a new light on an age-old mystery.

16

Philosophical Perpetual Motion and Atomic Energy

There are some aspects of perpetual motion which you cannot see. In fact you cannot appreciate them at all as being perpetual motion. To the learned, though, they exist. It is, you might say, all in the mind! Take atoms, for instance. Now there's perpetual motion of particles on so grand a scale as to be total.

Perpetuum mobile, and its Italian equivalent, *moto perpetuo*, were terms introduced into music to describe a rapid instrumental composition characterised by the notes all having the same value. Weber, Mendelssohn, Paganini and others have used the term.

Again, there are the equivoques in *perpetuum mobile*. One of the best known is the theoretical overbalancing wheel which would work. Take a disc of cardboard and from its centre draw, say, seven or nine radial lines like spokes. Stick a pin through the centre and while turning the disc with one hand, write the number 6 at the end of each spoke as it comes horizontal to your writing hand. When you have put a 6 at the end of each spoke, prepare what the student of mechanics would call a diagram of moments about the centre. The 6s on the right side add up to one figure but those on the other side have become 9s, and when you add these up, the number is higher than that on the right. The wheel should then rotate anti-clockwise to equiponderate. However, as it turns, each 6 becomes a 9 and the moment of all the 9s is always greater than that of the 6s. This means that the disc will spin, for you have created perpetual motion—in theory! (See Fig. 91.)

From absurdities like this, it is time to look at the real *perpetuum mobile* in miniature. A very interesting phenomenon is that known as Brownian movement or *pedesis*. It was discovered about 1827

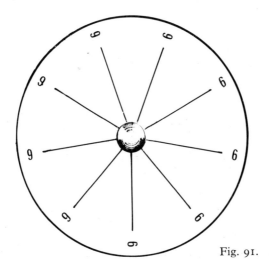

Fig. 91.

by a Dr Robert Brown (1773–1858) who was examining the minute contents of pollen grains suspended in a volatile liquid. Through the microscope he was astonished to see that the tiny particles were in a state of continual agitation 'looking for all the world as if they were alive'. An interesting way of observing this is to rub up a little gamboge (gum resin) and deposit it in water. Immediately it breaks up into exceedingly small globules. If a drop of this is now viewed under the microscope, the gamboge globules will clearly be seen to be moving at great speed and in all directions.

The explanation is that the movement is due to molecular bombardment and the impact of invisible molecules against the globules sets them in motion. The effect is perpetual and once more Nature seems to poke fun at Man's puny attempts by exhibiting her own tiny perpetual motion machines to us.

These moving particles obviously display the release of energy but how it could ever be possible to harness this infinitesimal power I cannot imagine.

Another form of perpetual motion for which it is impossible to see any practical use is concerned with a strange electro-magnetic paradox discovered in the thirties. It was found that if a current was induced in a lead ring which was cooled to the temperature of liquid helium, the current would stay unchanged for an indefinite period, certainly for many years, just flowing round and round. Initially, forty years ago, the length of time which the current could be kept in unchanged motion was determined by how long laboratory conditions could maintain the low temperature. The record was about three years, at the end of which time the electrical current was still present in the lead ring and remained undiminished in amperage. The experiment, now a regular laboratory demonstration written into the textbooks, shows how the super-cooling of the lead ring reduces electrical resistance to nought and so produces a sort of perfect flywheel.[1]

There is one form of perpetual motion which is everywhere and upon which everything depends. This is the perpetual motion of the atom and, as every schoolboy physicist knows, absolutely everything is made of atoms. In simple terms, an atom is akin to a macroscopic solar system: it has a 'sun', termed a *nucleus*, and a number of 'planets' which are called *electrons*. Electrons and neutrons are about the same size, each being around a millionth of a millionth of an inch across.

The nucleus itself comprises two different types of particle, *protons* and *neutrons*. Protons carry a positive electric charge; neutrons are neutral and electrons carry a negative charge. For every positive proton there is a negatively charged electron. These component parts are in constant motion, but they are held together to form the whole (the atom) by electrostatic forces. An easy analogy here is the motion of the sun, the earth and the moon: all are separated, yet all maintain regular and symbiotic perpetual motion.

Energy can be released from substances in several ways. Take wood, for example. You can burn it. This serves to rearrange the electrostatic forces which hold together the atoms themselves and produces heat by what is termed *chemical energy*. It does not matter how careful or thorough you are in arranging the burning process, only a fraction of the available energy is released in this manner.

A much more thorough 'burn' can be undertaken by actually breaking open the nucleus itself—this used to be called 'splitting the atom'—to produce *nuclear energy* in the form of heat. The advantage of nuclear energy over chemical energy is at once appreciated when you discover that for the same weight of fuel, nuclear energy can release in excess of a million times as much heat as chemical energy.

[1] Those seeking much fuller details of this experiment should consult the paper written by P. Grassmann entitled 'Untersuchungen über die Mikrowiderstände der Supraleiter' published in volume 37 of the *Physikalische Zeitschrift*, 15 August 1936, pp. 569–78. This phenomenon is called *super-conductivity*.

This simple-sounding system with its obvious advantages led to the perfection of the nuclear power reactor in which certain atoms can be bombarded with neutrons until, rather like a single sperm fertilising an egg, one neutron gets through to the central nucleus. When this happens, the central nucleus shatters into fragments resulting in the release of energy in the form of heat. The shattered central nucleus is now in the form of separate neutrons and these can be accelerated so that they can bombard more atoms. This is the process known as *chain reaction*. The end product is electricity converted from heat.

So far, so good. The chain reaction is not, as was devoutly feared in the thirties, a terrible form of perpetual motion of destruction which, once begun, would not stop until the entire universe had been converted into heat. It can be controlled by absorbing the flow of excess neutrons in a heavy screening of a suitable material, rather in the way that an earth bank behind a rifle range prevents spent rounds from perforating passing buses or stray cattle. But this is not achieved without something unusual taking place. This packing material becomes changed into another substance. Unlike the alchemist's dream, though, it does not become gold. It becomes fresh fuel for the reactor. If the Second Law of thermodynamics is not exactly shattered by this, then it has certainly become a little bent.

What really happens is worth taking a minute or two to explain because it does savour of the creation of matter. I emphasise that I am not going to bore you with too much nuclear science, but just enough to illustrate the unusual circumstances of what is termed the *breeder reactor* which manufactures more fuel than it uses. After all, that sort of ability would have been cherished by the Keelys and the Willises and the Redheffers a hundred years ago.

Nuclear breeding is achieved with the neutrons released by the process called nuclear fission. Fission really means splitting, and the fissioning of each atom of a nuclear fuel such as the natural substance uranium 235 releases an average of two-and-a-half high-energy neutrons. In order to keep the reaction going continually (a chain reaction), one of the neutrons must trigger another fission and the rest are divided between those that are lost and those which are available to breed new fissile atoms. This means that, in nuclear jargon, they can transform 'fertile' isotopes of the heavy elements into fissile isotopes. There are several of these fertile raw materials. One is thorium 232 which will change into uranium 233 when united with a virile neutron, and another is uranium 238 which becomes plutonium 239. You will note that uranium 235 and uranium 238 along with thorium 232 are natural substances whereas uranium 233 and plutonium 239 are artificial.

Breeding (the making of fresh fuel) occurs when more fissile material is produced than is consumed and the quantitative measure of the process is the doubling time. This is the length of time required to produce as much net additional fissile material as was originally present in the reactor. At the end of the doubling time the reactor will have produced enough fissile material to refuel itself and to fuel another identical reactor. An efficient breeder reactor can be expected to have a doubling time of between seven and ten years.

There are two types of breeder reactor, qualified by the material which they use and the substances transmuted, i.e. changed into something else. The thermal breeder makes use of slow neutrons and probably operates best by breaking up thorium 232 and turning it into uranium 233. This is usually called the thorium cycle. The more efficient fast breeder employs more energetic neutrons and is best fuelled with uranium 238, resulting in plutonium 239. This is called the uranium cycle. Fast breeders are more efficient than thermal breeders because the non-productive absorption of neutrons is less than in thermal reactors, and the consequence is that the doubling time is speeded up.

From the world economic point of view, breeder reactors are invaluable as power producers for they make it possible to make use of enormous quantities of low-grade uranium and thorium ores dispersed in the rocks of the earth and this could be a source of low-cost energy for many thousands of years.

To quantify the capability of the breeder with its vast capability for releasing the energy stored in those strange minerals uranium and thorium, over the next fifty years the use of breeders as at present planned can be expected to reduce by 1·2 million tons the amount of uranium which would have been consumed to produce electricity without breeders. This is equivalent to three thousand million tons of coal, give or take a bucketful.

Whereas the thermal reactor has an efficiency of about one per cent of the potential, the breeder is up to between 50 and 60 per cent. This strange statistic becomes clear when you realise that if it were possible to fission all the atoms in a pound weight of natural uranium it would release as much energy as three million pounds weight of coal. Thus even the dubious one per cent of this is a valuable end-product. The efficiency sounds on the low side because, with present systems, the fission products in the form of atomic ash accumulate and slow down the process of fission.

If we could completely fission all the atoms in one gram of uranium 235, it would be possible to produce electrical power to the extent of one megawatt day or thereabouts. In other words, fissioning one gram of uranium 235 a day would produce about one megawatt. At present efficiency, though, from one pound weight of natural uranium it is possible to produce as much electricity as a power station would generate from 10,000 lb. of coal—about 3,000 megawatt days/tonne.

When you see black smoke trailing behind a jet-engined airliner, this is largely unburned fuel products and so this balance of productive fuel is wasted. If the smoke could be recirculated through another combustion cycle, more of the energy in the fuel could be removed. This, though, is not a practical proposition for an aircraft engine. For the nuclear reactor, however, its low efficiency does not mean that the unfissioned material is wasted. The practical solution exists to burn it again and again until just about the last feasible ounce of energy has been released and nothing remains.

Supposing 1,000 lb. of natural uranium is taken as a start and this is then separated into its two kinds of atom, giving us 7 lb. of uranium 235 and 993 lb. of uranium 238. By starting nuclear combustion in the 7 lb. of 235, for every atom of the 235 that fissions, an average of $2\frac{1}{2}$ new neutrons are produced. One carries on the fission as already explained, and $1\frac{1}{2}$ are spare. Now if the 993 lb. of uranium 238 are packed round the fire of uranium 235, the $1\frac{1}{2}$ spare neutrons pass into the 238 and produce $1\frac{1}{2}$ atoms of plutonium for each atom of uranium 235 that fissions.

When all the 7 lb. of uranium 235 has been consumed, there remains a mass of $982\frac{1}{2}$ lb. of uranium 238 containing $10\frac{1}{2}$ lb. of plutonium. Separating the plutonium from the uranium, we can now start a fresh nuclear combustion in the $10\frac{1}{2}$ lb. of plutonium, stack round the fire the $982\frac{1}{2}$ lb. of uranium 238 and produce a further 16 lb. of plutonium. The process can be continued in this manner until all the original 1,000 lb. of natural uranium has been fissioned.

What the process is doing is converting a product of combustion into a quantity of fresh combustible material. The breeder reactor is an infinitely more effective converter of natural material into energy than any other means and it promises to become even more efficient as the years go by.

In practice, there still has to be a screen of impure natural uranium of certain proportions to help contain the radiation, so extra uranium 238 may be added. At the end of the doubling time, when more plutonium 239 exists, this is used to fuel another reactor. Were it not for the ever-increasing demand for electricity which is likely to accelerate for many decades to come, we might soon find ourselves with a growing stock of nuclear fuel which we could not burn fast enough. Fortunately there is no chance of this happening for a number of reasons. The contribution to the National Grid from nuclear-generated electricity is still very small. In time it will probably replace most of the other generating methods such as fuelling by coal and oil. Reactors built now have the future demands in sight and the fuel they produce will finally fuel reactors not yet built.

Perpetual motion? Alchemy? The philosophers' stone? Nuclear science is far and away removed from such notions but it poses some interesting challenges which can be viewed in the light of whatever interpretation we choose to place upon the word 'perpetual'. Natural radio-activity is a strange and awesome thing, as shown in Chapter 12. Nuclear reaction can produce new elements —materials quite foreign to Nature. Radio-activity is quantified by the length of time it takes a substance to lose half of its strength or power. Man-made plutonium 239 will continue to radiate activity for a long time. Its half-life is 24,000 years.

Our whole world—everything in it—is composed of atoms, each being composed of particles in constant motion. Even the most solid thing you can think of—a block of steel, a cathedral or a battleship—is made of neutrons in perpetual motion. In accomplishing the astonishingly complex task of fissioning atoms, Man has succeeded in what was once considered the superhuman task of actually stopping for ever the perpetual motion of some neutrons around their original protons.

17

The Perpetuity of the Perpetual Motion Inventor

In Chapter 13 I referred to the dwindling number of questors for perpetual motion. Although their numbers are diminishing steadily, I cannot imagine that the perpetual motionist will ever really die out. The problem is that today the perpetual motionist may not realise that he is seeking perpetual motion, but may be convinced that he is on some other sensational course, while our own training advises us that what he is really doing is trying to meddle in perpetual motion. In a moment I will give an example of this, but first let me conclude my argument. As I have said, unintentionally or otherwise people will always seek something for nothing. In an age when we can sit at home and watch our fellow men walking on the moon, when we can see nuclear reactors producing energy, when the flick of switch can set in motion a whole string of labour-saving machines in the home, perpetual motion appears to be knocked right down to size as just another simple mechanical or electronic operation. This conditioning of man has come about partly through the environmental substances around him which I have already referred to. It also stems from the mass attitude to mechanics which means that, although our potential level of intellect is far higher today than, say, fifty years ago, our understanding of things is frequently far, far less. At the end of the last century, almost every township of any size had its own Mechanics' Institute where the working man with an inquiring mind could—and did—attend lectures and demonstrations on science, electricity and mechanics.

Today, the average person leaving school seems to leave behind him all desire ever again to touch a serious book, to have to think productively in the arts and sciences, and to create, to write, to experiment. Naturally we are far, far better off today than even fifty years ago. Our society is less class-structured, the state provides that we shall not go hungry and we have comforts for our bodies. In place of the Church we have other, less disturbing calls on our souls. The Lord no longer has to watch over us—it is a task taken over by various ministry departments. We are better fed, better housed, we have better artificial light (have you ever tried to work by gas-lamp?), better roads and potentially better public transport. The price we have paid for all this is the will to make our own entertainment and to amuse ourselves with whatever hobby takes our fancy. We have been schooled out of the right to make fools of ourselves if we wish and make perpetual motion machines or whathaveyou. Personally, I am sorry that there seems to be so little left for the ordinary man to invent, to experiment with, or to gain wonderment from. Of course there are still those who sublimate their otherwise repressed urges by collecting, by thinking and by writing. But they are relatively few in number. The psychiatrist calls it escapism. I think this is too strong a word and those who collect musical boxes, magic lanterns and stamps, or those who spend their free time standing waist-deep in a trout stream, climbing mountains or jumping out of aircraft with a parachute are in truth giving themselves personal pleasure in a manner which should not upset others.

But I have veered off the subject of perpetual motion and one thing which I made mention of in Chapter 3 was a remarkable hark back to the primitive attempts at perpetual motion which took place in Ohio rather more than a century ago. This was related by William Marion Miller of Miami University, Oxford, Ohio,[1] to whom I am indebted for the following account.

The events about to be related took place in a little hamlet named Maudville, also known locally as Mauds or Maud's Station, in the south-eastern part of Butler County, Ohio. What happened here could be described as a monumental folly, and yet few people are even aware of the events which took place there late in the 1860s or early 1870s. The only place in which these events seem to have been committed to print was the *History of Butler County*, published in 1882 by the Biographical Publishing Co. of Cincinnati, Ohio and even this was but a brief description. An examina-

[1]*Isis*, vol. 37, pt. 1–2, 1947, p. 57.

tion of the county newspaper for the period concerned was undertaken by Mr Miller, and he found no mention of it there. He found that it was recorded only in the memories of a few old people who had heard of it in their youth, and nobody is alive today who witnessed it. One more negative feature must be mentioned: even the name of the man who conceived the scheme has not been recorded.

The story is that this man decided to build a sawmill along the right-of-way of the newly built railroad then known as the 'Short Line' (i.e. the shortest line between Cincinnati and Dayton, Ohio) and now part of the New York Central System. This was to be no ordinary sawmill, for water-power was not available for the conventional type of mill and the mill-owner chose not to use steam. Perpetual motion experiments were, at that time, numerous and so our nameless entrepreneur thought up a method of harnessing for practical use this seemingly inexhaustible force as the motive power of his new mill. No experiments were made in advance of construction work, no working models were made and all thinking was on the theoretical side. As is now obvious, he possessed not the slightest notion of the simplest laws of physics and, apparently, he would not listen to those who had.

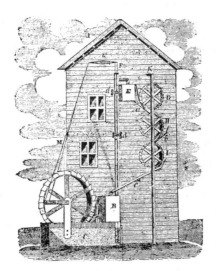

Fig. 92. There are periodic recurrences of attempts to produce perpetual mills. This one was sketched less than a century ago.

The would-be capitalist (old people assured Miller that he planned to expand his 'discovery' and quickly become wealthy) constructed a sawmill with a unique 'power plant'. He first built a huge vat, mounted on heavy supports and some fifteen feet in the air. This held about a hundred barrels of water and he and his family filled it up by hand. Then he constructed a trough or race down through which the water ran, dropping over a conventional waterwheel *en route*. Finally the water spilled into a much smaller cask. The wheel, so the builder firmly believed, would spin round under the force of the water and rotate a pulley connected by belts and levers to a pump and to a saw. The water would be immediately pumped back to the upper vat and soon start down once more to perform its useful task. The saw, meanwhile, would operate merrily.

Theoretically, thought its designer, once the cycle was started here would be a source of free power that would never come to an end. Admittedly a gallon or two might be lost through evaporation and would have to be replaced now and again, and presumably the system would not operate during freezing weather, but aside from that it could be kept in operation for at least three hundred days of the year if the owner so desired.

Now as earlier stated no experiments had been made, nor had the precaution of the construction of working models been undertaken beforehand. Nevertheless, the builder was so convinced of the workability of his scheme that he built the entire mill, purchased a large supply of logs from the surrounding farmers, hired his staff, and announced the day on which he would begin operations.

The local history states (and stories told to Mr Miller around 1917 by old people who had been young at the time corroborate the account) that a large crowd assembled to watch the enterprise get under way. The upper vat had been filled previously, and when the valve allowing the water to escape was opened, the water started on its first supposedly round trip. The wheel turned, but little else happened. Amid the guffaws of the watchers, the water poured over the sides of the lower cask—and the attempt to harness a cheap and never-failing source of power quickly came to an unglorious end.

As far as W. M. Miller has been able to trace, the owner-inventor experimented no further but immediately abandoned his enterprise and devoted himself to other pursuits elsewhere. The mill, casks, wheel, logs and all soon rotted to the ground and were obliterated in the course of a few years. Even memory of it was soon lost.

No drawing seems to have been made of this unique sawmill, and nothing more than local importance appears to have been accorded it.[1] It served, though, to epitomise the American quest

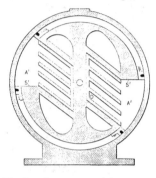

Fig. 93. Compressed air perpetual motion machines range from the patently stupid to deceptively simple yet well-engineered items like this one. Compressed air (or other elastic fluid) is contained in the chambers A1 and A2. Because the surfaces S1 and S2 are larger than the others, the reaction on these is supposed to produce a rotary motion. No air is exhausted but should any escape past the seals, the motor also drives pumps to replenish the air pressure.

Fig. 94. Dating from 1902, this motor was claimed to depend upon a revised statement of the First and Second Laws of thermodynamics and to consist of a special cycle of working in which the heat rejected in the Carnot cycle was interrupted and made to return to source so making it possible to convert into motive power the diffused heat at ordinary temperatures that exists in the atmosphere or elsewhere. The vessels *b*, *c*, *d* and *e* are mounted on a shaft *a*, and have one side *f* tangential to the shaft, and the other side radial. Compressed air is forced into each vessel through the valves *p*. It is stated that 'under the action of the internal pressure of the vessels, and after a slight impulse is given to same, in the direction of the arrow, the whole apparatus will begin to move and continue to do so without ever stopping, the velocity corresponding to the pressure established within the vessels'.

[1] Cf. above, p. 209.

Fig. 95. Clockwork has always been a simple way of driving mechanisms but the perpetual motionist has long tried to find a way of winding his weight or spring up again automatically. James Cox succeeded admirably with his clock. Not so the inventor of this system who suggested that trams and cars, sewing machines and 'other light machinery' might be clockwork-powered on the perpetual motion principle. His spring, contained in the centre at *A*, drove its own winding spindle through an arrangement of bevel gears, compound wheels and pinions. And, presumably, still had power to drive a tram!

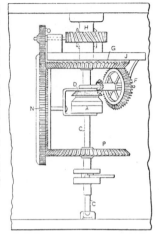

for perpetual motion as a motor for practical use which raged with such fervour around the middle of the last century. Even at that time, its misguided technology was two centuries old. The water-wheel which shall feed its own mill-stream, as we have seen in earlier pages, is probably as old as the first miller who experienced the problems associated with the flow of water in a dry season. Bishop Wilkins showed one form of this in his *Mathematical Magick*, using an Archimedean screw to raise the water again. And at least 150 years before our Ohio mill-owner thought of it, a plethora of projects had been drawn up which represented the water-wheel minus the screw, the water being pumped back up again in its own buckets.

If one generalisation has to be made, then it must be that all perpetual motion seekers and inventors were sanguine and possessed of an enthusiasm the quality of which does them credit.

Some people thought up wonderful schemes which, although they did not realise it, were nothing but perpetual motion machines. Consider the American lady from New England who in 1929 invented a cylindrical washing machine which had no motive power other than the falling of the clothes from the top to the bottom of its divisions. Continually—of course ...

Earlier in this book, I promised to relate the strange tale of a perpetual motion-type machine with which I became involved.

Some years after I came out of the Royal Air Force, I founded a small company to market a light aircraft which I had designed.

As designer/director of this business, I certainly never made anything like a fortune and in fact over the years the then underdevel-
oped private aircraft market helped me to lose money. But we
did produce several successful light aircraft designs. One of these
was a very small single-seat ultra-light sporting monoplane which,
for the purposes of this story, I will call the Finch. A tiny aeroplane, it could be stored, wings folded, in a garage, and would
cruise at around 85 miles per hour on 37 horsepower provided
by a flat twin-cylinder engine. From this you will at once gather
that this was no mammoth doyen of the skies but designed so that
those who wanted to fly could do so very cheaply, simply and
safely. The little aeroplane received a good amount of publicity
in technical papers and the 'dailies' and this resulted in a plenitude
of correspondence, very largely uncommercial and emanating for
the major part from youthful aircraft enthusiasts who just wrote
to ask odd questions, or from those who wanted to know if we
could modify the aircraft as a biplane, as a five-seater, or to fit
seaplane floats to it.

One day in the autumn of 1958 I received a strange letter from
a man in North London. It did not take me long to realise that
he was, to put it mildly, possessed of unusual reasoning powers.
His letter read:

'I am interested in your plane ... I will get to the point as to the
reason for my interest [in the Finch] and I hope I will succeed
through this introductory note, when you have learnt its contents,
in exciting in you an equal interest. I want one of your planes,
and I wish to have installed into it an engine of mine that I have
invented. This engine I have reasonably developed to a practical
stage. There is no fuel required ... (please do not, upon reading
the latter, allow unworthy discreditable thoughts to arise in your
mind before you have an understanding of it), which is the most
natural, silent, smooth running engine in the world, and the engine
of the future ... With this engine, which will be a light one to suit
the "Finch", this plane should be able to attain quite easily a speed
of between 300 and 400 mph. My only disturbing thoughts are
as to whether the wings, etc, will stand the strain. With some practice, after a performance between 300 and 400 an hour, and staying
in the air about ten hours, flying around England and the Channel
Islands, I will learn its weaknesses, which you could then have

strengthened for me. And then I intend crossing the Atlantic with it ... It will then be the most wanted engine in the world.'

Naturally, my correspondent looked for the enthusiastic co-operation of my little company inasmuch as I was to supply the aeroplane, free of charge, on the promise of the enormous publicity which such a flight would gain. The Finch, I might add, was designed for a maximum speed of 100 m.p.h. and, were it to attempt flight at 400 m.p.h. (assuming that the projected engine could overcome the high drag of a low-speed airframe at that speed), I could quite see that a ten-hour preparatory flight would be ample to reveal its structural weaknesses. Like the wings coming off as the 150 m.p.h. mark approached.

It was a busy time for my little company and Mr X's letter remained unanswered for about a fortnight. Then came a second letter begging for an answer as to whether we would co-operate. I then proffered a rather formal reply, telling him about the designed maximum speed of the Finch, and quoting him an ex-works price for the complete aircraft. I also stated the airworthiness requirements which would have to be satisfied before any new engine could even be considered for installation.

Because curiosity is a trait which has never left me since I was a schoolboy, and since it so happened that Mr X lived within a very short distance of my then London address, I called at his home on the spur of the moment late one afternoon. It was in one of those once grand but now rather mouldering Georgian terrace streets running off the Finchley Road on what the Hampstead elite politely refer to as 'the wong side'. The street, at the crossroads between restoration and redevelopment, was now part of the receding bedsitter-land of North London and each front door marked several people's off-the-peg castle. The roadway was cluttered with aged cars. I found the house but Mr X was not in. I poked my card through the letter-box.

My impromptu visit gave to Mr X rather more encouragement than I would have wished and a flurry of letters from him followed. He urged that we should get to know each other since we would have to work together if he was to arrive on the other side of the Atlantic in one of my Finches (which normally had a maximum fine-weather range of about 250 miles). He added that an aircraft fitted with his engine 'will be dependent on wind or gravity pull (my engine will increase the power of the two latter by about 40 times)'. Wondering how on earth an aircraft powered by an engine which increased the power of gravity 40 times could do anything other than fall through the ground, and feeling that I was out of my depth already, I arranged to visit him.

I found a small man in his sixties, fast-talking, enthusiastic and of great artistic bent. His flat was cluttered with old furniture and artists' requisites and upon the walls were some of the most remarkable portraits in pencil which I have ever seen. This man certainly did have talent in that direction at least.

The upshot of our meeting was confused to say the least. This man had never even been in an aeroplane in his life, let alone piloted one himself. He had no idea of flying, but believed that the only way to learn was to get into an aeroplane and fly it. I chuntered on about the navigation problems of flying a tiny aircraft like the Finch such a great distance as an Atlantic crossing would involve. I told him some of the legal aspects of pilot's licences, the need for flying instruction and the insurance problems. He would have none of it. He was convinced that he could fly without interference from an instructor.

I then asked about his engine. Yes, he had built one and it used no fuel but worked on a new principle. Where, I asked, was it and might I see it? He explained that he had stood it on the floor in the hall but, since he was unable to stop it, its noise had annoyed the other tenants, and after several weeks of incessant operation he had been forced to dismantle it in order to restore the peace of mind of his landlord. Where, I asked, were the pieces that I might have a look at them? Mr X told me that they were distributed about his flat, all were inaccessible at that moment and in any case he wanted an assurance of my integrity before he could confide in me. Once that confidence was established, he said, he would reassemble the engine and tell me just how it worked.

He stressed that his engine would 'positively increase the power of all motors and engines (without adding any extra fuel), at least twenty times (not per cent—*twenty times*)'. For my part, I had to guarantee to provide an aircraft and full back-up for his flight before I could be let into the secret. 'The risk', he kept saying, 'is all on my side.' I did not see it this way and, since the Finch had never been responsible for hurting anybody, I did not choose to take action which might blot its happy reputation.

A few letters followed, but the matter, as far as I was then concerned, was concluded. Recently, in connection with this book, I tried to trace Mr X. The house is still there, just rather more decayed, but he has gone without trace. I still have all the correspondence, kept for its curiosity value, and I often wonder what happened to his fine pictures.

When, a few years ago, an American announced his intentions of flying around the world non-stop in a light aircraft, this might have been seen as a variation on a similar theme. In fact it was not. The pilot in question, an outstanding aviator and engineer, had proved that under ideal conditions, and using a specially modified aircraft and engine, it was quite possible. At the time of writing he has yet to make the flight, but I believe that this can succeed. Without bending the laws of mechanics or thermodynamics, or evoking the assistance (other than the blessing) of the Almighty.

Reverting to perpetual motion, we have seen that the perpetual motion seekers were of several types—the creative who became disheartened, the visionary who theorised, and the trickster who could separate fools from their money in double-quick time. In more recent times, perpetual motion has become a scientific possibility in certain aspects and the perpetual motionist, no longer tainted with avarice and, usually, better equipped with knowledge and technology than his forebears, has become a dimorphous creature. On the one hand, we find the nuclear researcher and the space scientist, and on the other somebody who sees a legitimate commercial proposition in an intriguing laboratory experiment.

A strange and quite fascinating toy appeared in England in 1948 which was a clever attempt at perpetual motion in its simplest, most basic form. The overbalancing effect of a tube of mercury and a small weight, although in no way perpetual motion, was understood from earliest times. In 1885, a tiny toy gained extensive popularity in Paris, and even this was but a rekindling of an old, old idea. Two tiny puppets stood one behind the other and apparently holding two long tubes. The tubes contained a little mercury and the puppets were pivoted to them. When placed on the top of a staircase, the toy slowly and decorously 'walked' downstairs, each doll leap-frogging over the other as the mercury moved up and down the tube. Another, much simpler toy with the same effect is the long floppy coil spring which 'walks' down a flight of stairs.

The 1948 toy (and I believe that it is still made today) was extremely ingenious. It was dubbed the Drinking Duck, and when it appeared it proved probably the most controversial thing of those immediate post-war years. Shopkeepers placed them in their windows—and were sure of a crowd forming to watch it.

The Drinking Duck consisted of a metal stand shaped in the form of a pair of legs. The duck's body was a glass tube with a thermometer-type bulb at the lower end and the top formed into a duck-like beak. The tube was painted and decorated to look like a caricature of the familiar bird and it was pivoted to the legs so that it could swing freely. When placed so that it confronted a glass of water, the normal position was for the duck to assume an upright position. However, once the beak had been dipped in the water, a process began which, if not perpetual, could certainly be described as *extended* motion, for the duck would systematically dip its bill in the water, swing back upright, and then gradually dip forward, immerse its beak, then stand back up again for the process to continue over again.

The secret lies in the body of the duck which contains 'a volatile liquid' which normally rests in the lower portion of the tube and in the bulb. This liquid evaporates under room temperature, and when the head of the duck is first cooled by the evaporation of water on it, the vapour inside the head condenses, reducing the pressure and so allowing the vapour pressure in the lower end to raise the liquid in the neck high enough to overbalance the head, causing the duck to dip down to 'drink'. The vapour pressure in the lower end is then released, allowing the liquid to return by gravity and the duck swings back to repeat the process. When properly set up with the glass of water at the correct level, the operation becomes continuous. From the upright position, the duck slowly begins to lower its head until finally the bill touches the water when it immediately swings back upright. As might be expected, it is necessary for the temperature of the water to remain slightly below that of the atmosphere. This amusing and instructive toy used to sell for 11 s.

I mentioned just now coil spring toys. There is one spring device which might be dubbed the toy of the philosophical scientists, to use an eighteenth- and nineteenth-century term. This is the

Wilberforce pendulum which appears to be a probable candidate for the modern perpetual motion stakes. This device is a specially wrought coil spring suspended vertically and provided with a weight fixed at its lower end. Unlike the normal swinging pendulum, this one moves up and down and it puts into practice the fact that when a coil spring is stretched, a longitudinal force is exerted. Consequently a coupling exists between the two motions of the pendulum—vertical oscillation combined with alternating clockwise and anti-clockwise oscillation. As it stretches, the coils unwind a little, and as it contracts, they recoil. The pendulum is so adjusted that energy is being constantly transferred from one condition to the other and thus the extension of the pendulum (the alternations between vertical and rotary oscillation) remains constant because, disregarding the losses due to air friction and resistance, the sum of the energies remains constant. In a total vacuum, then, the Wilberforce pendulum might remain in motion for an extended period of time but, since a *total* vacuum is practically impossible, perpetual motion is once more denied.

The perfect concept of perpetual motion is a ball spinning between two magnets suspended in a vacuum, but for similar practical reasons this is a virtual impossibility. Again, were such a device to be made, how might its movement be applied as a motive power? The simple answer is that it just is not possible. A ball moving in a glass tube could be used to operate an electronic switching function and might even be able, as a function of an experimental mechanism, to produce an electrical current. To do any of these things, however, complex, costly and energy-fed mechanisms must be employed as well. In a laboratory it is an amusing experiment to show, for example, that measurable electrical discharge can be recorded by shining a light on to a light-sensitive cell, but the input of energy is always greater than the output. Motion, Nature decreed at the outset of time, shall not be given away for free.

18

A Summing Up

The difficult, the dangerous, and the impossible have always had a strange fascination for the human mind.

It is with this succinct observation that John Phin opened his interesting little book *The Seven Follies of Science* back in 1913. He defined the follies as:

1. The Quadrature of the Circle or, as it is better understood, the Squaring of the Circle
2. The Duplication of the Cube
3. The Trisection of an Angle
4. Perpetual Motion
5. The Transmutation of the Metals (Alchemy)
6. The Fixation of Mercury
7. The Elixir of Life

To those of us today who read this, most of these sound like non-events, yet in truth their total fulfilment or accurate completion remain largely impossible. Strange to say, of all these follies, there is only one which still excites us today and this is one to which a great amount of research and money is still devoted both by the crank and by the studious research institute. This is the Elixir of Life or, as it is now termed in the less ebullient language of the twentieth century, the prolongation of natural life. Monkey glands and herbal treatment have given way to sophisticated techniques involving such diverse approaches as drugs, freezing and component transplants.

The rest of the follies mean nothing to us any more. Besides a few devoted experimenters, call them cranks if you life, Man has forgotten all about Perpetual Motion, Alchemy and strange things like Palingenesy (a sought-after process that would regenerate a living thing from the ashes of the dead) and the Fourth Dimension (you had best contrive your own explanation of that one since I cannot visualise it). Man has moved into the realms of practicality and systematically inured himself to those things which aroused excitement and wonder in past generations. This is in many ways a great pity since many of Man's abilities have been made redundant over the years and he no longer possesses the powers upon which his antecedents placed great reliance. Only the country dweller, for instance, can now observe and relate to the signs of Nature and weather in a manner which can prove beneficial to his way of living. Natural skills and senses are of little use in the concrete world of hustle and bustle in which most of us live. With this change which is almost evolutionary (although it has been brought about in, relatively speaking, a remarkably short period of time), Man the inventor, Man the questor, has been replaced by a different form of inventor and questor. Where once Man worked with definable tools in a small way, he now works with indefinables on a vast canvas which can embrace the whole world. Where once he thought with grey matter he now computes with grey boxes, and the problems which beset him bear no relationship to those which concerned mankind even half a century ago.

Beliefs have been shattered, the simple made wiser, the foolish generally made less foolish. The sentiments expressed in Exodus 35, verse 2, and adhered to by God-fearing generations, are now made to sound somewhat hollow, and what Our Lord ordained we should do on Sunday bears little relationship to twentieth-century Man, continuous-process industries, the personal call for wealth and our own choice of what Sunday is for.

Where does one place the inquiring man with an inquiring mind today? Nowhere, usually. Only the very few now pursue really creative private activities and, whereas a student group might, for an exercise, set a computer to trisect an angle to n decimal places recurring, or have a computerised drafting machine re-create isometric views of a cube, our goals have generally become smaller as our abilities have lessened. The pocket calculator, now used even in schools, means that we no longer need to learn how to do simple arithmetic. A whole new pattern of life has emerged in which living has become so easy that, unless we find something with which to exercise our brains, they stand a good chance of becoming atrophied. For most of us, the mass entertainment market must be relied upon to stimulate our minds. So often this stimulation falls miserably short of what we need until the point is reached where we no longer need it, or recognise the need for it. It is so easy just to sit down and do nothing, just watch entertainment. Even the family musical evenings around the piano or American organ have departed and family games are reserved for Christmastime and birthdays. Music comes via the radio or gramophone.

What has all this got to do with perpetual motion, you are asking. Well, I promised in the title of this chapter a general summing up and what I have done so far is to establish the fact that most of us today are easily moved to ridicule the ideals which captivated the minds of man in times past. I rather believe that, in spite of our better standards of living, a large number of our predecessors demonstrated a greater power of creativity than most of us today, and the mere fact that time has proved their notions to be unrealistic should not make us think any the less of them.

It has already been established that a good proportion of our basic knowledge of physics and mechanics come about as a result of the impossibility of perpetual motion having been discovered. That, if no other thing, can be said in favour of the perpetual motion seeker and must count for something.

So where do the perpetual motion seekers really stand today? The main incentive to perfect perpetual motion was to be found in those times when power called for the physical effort of Man or animal. With simple, clean, efficient power by way of electricity available effortlessly, even the most misguided of men can usually find something better to do than to seek perpetual motion. Mind you, with Man never more conscious of the dwindling power resources of earth, forcibly aware of the pollution concomitant with modern industry and transport, and becoming desperate to make amends for the thoughtlessness of generations, some perpetual motionists, rather like the last-remaining specimens of a dying species, may always be present. And their notions can be expected to differ little from those who have passed before them. For there is nothing really new under the sun, certainly not for the run-of-the-mill (pardon the pun) perpetual motion seeker.

I will tell you something else which I find interesting. Talk perpetual motion for a while to the ordinary person and, sooner or later, the chances are that he will come up with a scheme of his own. Needless to say, that scheme can be depended on not to work, but it proves that deep down inside us all there is a latent belief that such a thing is possible, even if for no other reason than that it would be nice to succeed where others have failed.

In the course of writing this book, I have spoken to many people ranging from non-mechanically minded people in diverse professions to highly qualified engineers. One man, a member of the latter category, laughed at me and said, 'Why do you want to write about the ridiculous for surely you know it is impossible!' Anyway, I showed him some drawings of abortive efforts (they are contained in these pages) and he went away looking amused. A week later he telephoned to say that, although it was quite impossible, he had schemed out a perpetual motion machine which would theoretically work! This sort of thing has happened several times which demonstrates to me that the subject is not really moribund. And I'll wager that some of my readers will amuse themselves with pencil and paper to try to better centuries of failure.

I must add a cautionary note that, however infectious the enthusiasm for the subject, all these hundreds of years of trying should alone prove that the chances of success are unlikely. True, James Cox did succeed with a mechanical barometric perpetual motion, but his machine failed because of one thing which he had forgotten—the building housing his masterpiece was not perpetual and his handiwork was immovable without first being emptied of its precious life-blood.

Please do not send me perpetual motion schemes and ask me (a) if they will work, or (b) why they won't work. Although as a design engineer I have built many things, contrived many mechanisms and patented a few, I have never tried to make a perpetual motion machine. It is beyond my ability and, I venture to suggest, beyond yours too.

From the honest, hard-working miller of the seventeenth century who tried sincerely to improve the service which he could provide to the community in which he lived and worked, through to those who sought personal gain by deception, perpetual motion has travelled a strange and erratic path. In contrast with the path of development in most other aspects of the philosophical arts and sciences wherein progress has devolved by a process of experimentation and the systematic discarding of unworkable schemes, perpetual motion has traversed the years in an uncoordinated, almost aimless course. Certainly it has doubled back on itself time and time again as successive generations of inventors have tried vainly to succeed by applying a well-tried and worthless approach. Indeed, you might say that not only did perpetual motion fail in its goal as an automatic, free power for all, but it spent its entire fecund years going nowhere. Those very few inventors who had something worthwhile to contribute (here I mean men like Cox who found a practical way of timing time perpetually) failed, in their very breaking of the ruling, narrow parameters which limited the comprehension of the majority of questors, to succeed in showing the way. One would expect the undoubted success of Cox to have inspired men to transcend the useless weight-shifting wheel and devote their enthusiastic energies to trying to make the barometer drive something else. This was not to be. There must be something in the make-up of the perpetual motionist which, while urging him on in his quest for the impossible, encourages him not to deviate from the well-trodden path to certain failure. Even the alchemist who had a far broader sphere of experimental potential before him—chemistry being infinitely less confined than mechanics—knew when he was beaten. Aside from the nuclear engineers who experiment beyond the accepted realms of possibility, I have never met an alchemist.

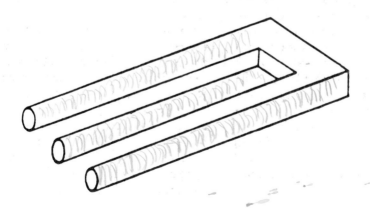

BIBLIOGRAPHY

ALLEXANDRE, Jacques (Father Dom), *Traité général des horloges*, Paris, 1734.

ANGRIST, Stanley W., 'Perpetual Motion Machines', *Scientific American*, New York, vol. 218, January 1968, pp. 114–22.

BAKER, Ray S., 'Tripler's Perpetual Motion', *McClure's Magazine*, New York, March 1899.

BEDINI, Silvio A., 'Perpetuum Mobile—the Invention of the Rolling Ball Clock', *Bulletin of the National Association of Watch & Clock Collectors*, New York, vol. 7, no. 6, February 1956, pp. 74–87.

BENHAM, Charles E., 'A Perpetual Clock—the Ingenuity of James Cox', *Scientific American Supplement*, New York, no. 1751, vol. 68, 2 October 1909.

'Perpetual Motion—the One Phenomenon which is Universal and Constant', *Scientific American Supplement*, New York, no. 2121, vol. 82, 26 August 1916, pp. 130–2.

BÖCKLER, Georg Andreas, *Theatrum Machinarum Novum*, Nürnberg, 1673 (reputed first edition), 1686 and 1703.

BRANCA, Giovanni, *Le Machine*, Rome, 1629.

BREWSTER, David, 'Perpetual Motion', *Edinburgh Encyclopaedia*, 1830.

CANBY, George, *Keely Motor Scraps* (collection of misc. and press cuttings), Franklin Institute, Philadelphia (shelfmark 531.8 qc16).

CARNOT, Nicholas L. S., *The Motive Power of Heat*, Paris, 1824. (reprinted 1912).

Century of the Franklin Institute of Philadelphia—1824–1924, Philadelphia, 1924.

CHAPUIS, Alfred and JAQUET, Eugene, *The History of the Self-Winding Watch*, Editions du Griffon, Neuchâtel, 1952.

CHARLESWORTH, F., 'Perpetual Motion Machines', *Cassier's Magazine*, vol. 29, December 1905, pp. 115–25.

COLDEN, Cadwallader D., *The Life of Robert Fulton*, New York, 1817.

COX, James, 'A Descriptive Inventory . . .', London, 1774.

DAUL, A., *Das Perpetuum Mobile*, Vienna, 1900.

DAY, J., article in the *American Journal of Science*, New Haven, 1850, vol. 60, p. 174.

DIRCKS, Henry C. E., *Perpetuum Mobile*, E. & F. Spon, London, vol. 1, 1861, vol. 2, 1870.

DYE, F., *Popular Engineering;* with chapters on Perpetual Motion, E. & F. Spon, London, 1895.

Encyclopaedia Britannica, 'Perpetual Motion', vol. 18, pp. 553–5, 9th edn, Edinburgh, 1883.

FRANCIS, G., *The Dictionary of the Arts, Sciences and Manufactures*, London, 1842.

GIBBS-SMITH, C. H., *Aviation*, Science Museum (HMSO), London, 1970.

GUILLEMIN, Amédée, *The Forces of Nature*, Macmillan, London, 1873.

HEINZE, Robert W., 'Why Perpetual Motion Won't Work', *Science Digest*, Chicago, vol. 24, August 1948, pp. 42–6 (abbreviated from *The Northwestern Engineer*, Evanstown, Illinois, March 1948).

Henry, Lord BROUGHAM, *The Circle of the Sciences*, London, c. 1870.

HERING, Daniel W., *Foibles and Fallacies of Science*, Van Nostrand, New York, 1924.

HEYL, Dr Paul R., 'Perpetual Motion in the Twentieth Century', *Scientific Monthly*, vol. 22, February 1926, pp. 143–5.

HICKS, Clifford B., 'Why They Won't Work', *American Heritage*, vol. 12, April 1961, pp. 78–85.

HISCOX, Gardner D., *Mechanical Movements, Powers, Devices and Appliances*, Munn & Co., New York, 1899.

Hobbies, 'Mechanical and Electrical Antiques', Chicago, vol. 52, February 1948, p. 28.

HOFFMANN, Franz, 'Die Perpetuum Mobile Theorie', Leipzig, 1912.

Hogg's Instructor, London, article in vol. 6, 1854, p. 278.

HORRINGTON, John Walker, 'Perennial Quest for Perpetual Motion', *Mentor*, New York, vol. 17, June 1929, pp. 49–52.

ISHERWOOD, B. F., 'Gamgee's Perpetual Motion Machine', *Kansas City Review*, vol., 5, 1882, pp. 86–9.

JANVIER, Antide, *Journal Encyclopaedique*, Paris, 1827.

JONES, T. P., article in the *Journal of the Franklin Institute*, Philadelphia, vol. 6, pp. 318, 414.

'Congreve's Perpetual Motion', *Journal of the Franklin Institute*, Philadelphia, vol. 7, p. 179.

KNIGHT, Edward H., *The Practical Dictionary of Mechanics*, Cassell, London, 1881–4.

LEBERT, J. A., *Du Movement Perpetual*, Paris, 1851.

LEUPOLD, Jacob, *Theatrum Machinarum Generale*, Leipzig, 1724.

LLOYD, Herbert Alan, *Some Outstanding Clocks over 700 years*, London, 1958.

MACDOUGALL, Curtis D., *Hoaxes*, 2nd edn, Dover Publications, New York, 1958.

MCKAY, J. T., article in *Galaxy*, New York, vol. 18, 1874, p. 625.

MACH, Ernst, 'The Science of Mechanics' (translated from the German by T. J. McCormack), London, 1893.

MILLER, William Marion, 'Attempts to Harness Perpetual Motion in Ohio in the Nineteenth Century', *Isis*, vol. 37, pt. 1–2, 1947.

MOORE, Clara J. Bloomfield, *Keely's Secrets* (docs., MSS), Franklin Institute, Philadelphia (rare books room, shelfmark M.78K).

MORITZEN, Julius, 'The Extraordinary Story of John Worrell Keely', *The Cosmopolitan*, vol. 26, no 6, April 1899.

MORRIS, Wilson C., 'Negation of Perpetual Motion in Elementary

Physics', *School Science and Mathematics*, vol. 13, June 1913, pp. 469–479.

MORTON, Henry B., 'Engineering Fallacies', *Cassier's Magazine*, vol. 7, January 1895, pp. 200–10.

'The Redheffer Perpetual Motion Machine', *Journal of the Franklin Institute*, Philadelphia, vol. 139, 1894–5, pp. 246–51.

NICHOLSON, William, article on Cox's clock, *Philosophical Journal*, London, vol. 1, 1799, p. 375.

ORD-HUME, Arthur W. J. G., *Clockwork Music*, Allen & Unwin, London, 1973.

'Perpetual Motion', *Engineering*, London, July 1972.

PAQUIN, R. E., 'Perpetual Motion Just Isn't', *Popular Mechanics*, Chicago, January 1954, pp. 108–11.

Penny Magazine, vol. 3, January 1834, p. 2.

PEPPER, J. H., *Cyclopaedic Science Simplified*, Warne, London, c. 1896.

PERKINS, F. B., article in *Galaxy*, New York, vol. 12, 1871, p. 341.

PHIN, John, *The Seven Follies of Science*, 3rd edn, Constable, London, 1913.

POLE, B. C., *Poleforcia*, London, 1899–1900 (Franklin Institute, Philadelphia, shelfmark 621.1 St.31.4).

Popular Mechanics, 'World's Greatest Hoax Recalled by Models', vol. 54, December 1930, p. 986.

Popular Science Monthly, 'Water as Fuel', March 1880.

Public Ledger Almanac (article on Keely's motor), Philadelphia, 1900, pp. 101–3.

ROBERTSON, J. Drummond, *The Evolution of Clockwork*, Cassell, London, 1931.

ROGET, Peter Mark, *Galvanism*, London, 1832.

ROUTLEDGE, Robert, *A Popular History of Science*, 3rd edn, Routledge, London, 1894.

SALEDINI, Valerii, *Perpetuum Mobile*,

SCHARF, J. T., and WESTCOTT, Thompson, *History of Philadelphia, 1609–1884*, Everts & Co., Philadelphia, 1884.

SCHOTT, Caspar, *Technica Curiosa*, Nürnberg, 1664.

Scientific American, 'A New Patent Office Ruling on Perpetual Motion Machines', vol. 105, 16 December 1911, p. 561.

'A Perpetual Motion Machine Problem' (R. P. Horton's submission), vol. 103, 17 September 1910, p. 214.

'Garabed', vol. 119, 31 August 1918, p. 162.

'Inventors of Perpetual Motion Machines', vol. 64, 23 May 1891, p. 238.

'Is Perpetual Motion Possible?', vol. 67, 30 July 1892, p. 64.

'Mediaeval Machinery—How the Germans of the Seventeenth Century Tried to Solve an Age-old Problem', vol. 115, 15 July 1916, p. 64.

'Perpetual Motion—Some Examples of Misguided Ingenuity', vol. 105, 18 November 1911, p. 452.

'Perpetual Motion Again' (G. W. Francis' submission), vol. 77, 13 November 1897, p. 311.

'Solutions of Mr. McNeill's Perpetual Motion Problem', vol. 103, 24 December 1910, p. 503.

'Something for Nothing' (Garabed), vol. 119, 13 July 1918, p. 26.

'The Perpetual Motion Problem' (solutions to Horton), vol. 103, 1 October 1910, p. 255, and 8 October 1910, p. 275.

'The Problem in Perpetual Motion' (Dennis McNeill), 26 November 1910, p. 422.

'Typical Perpetual Motion Fraud' (J. M. Aldrich's machine), vol. 81, 1 July 1899, pp. 1 and 9.

Scientific American Supplement, 'A New Radium Perpetual Motion Machine', no. 1866, vol. 72, 7 October 1911, p. 239.

SCOTT, E. A., 'The Keely Motor', *Proceedings of the Engineers' Club of Philadelphia*, vol. 14, no. 4, January 1898.

SEABORG, Glenn T., and BLOOM, Justin L., 'Fast Breeder Reactors', *Scientific American*, vol. 223, no. 5, November 1970.

SELLERS, Prof. Coleman, 'The Redheffer Perpetual Motion Machine', *Cassier's Magazine*, vol. 8, January 1895, pp. 523–7.

SHAW, Prof. Hele, 'Perpetual Motion', *Littell's Living Age*, vol. 176, 1887–8, pp. 376–9.

'Perpetual Motion', *Nature*, vol. 37, 12 January 1888, pp. 254 ff.

SMITH, James, *The Mechanic or Compendium of Practical Inventions*, London, 1825.

SOMERSET, Edward (Marquis of Worcester), *Century of Inventions*, London, 1655.

STEWART, Alec Thompson, *Perpetual Motion, Electrons and Atoms in Crystals*, Science Study Series, Anchor Books, Garden City, New York, 1965.

TAISNIER, Johannes, *Opusculum Perpetua Memoria Dignissimum*, Coloniae, 1562.

TALLMADGE, G. K., 'Perpetual Motion Machine of Mark Antony Zimara', *Isis*, vol. 33, no. 8, March 1941, pp. 8–14.

TOMLINSON, Charles, *Cyclopaedia of Useful Arts and Manufactures*, Virtue, London, 1854.

TYMME, Thomas, *A Dialogue ...*, London, 1612.

WAINWRIGHT, Jacob T., *Principle & Mathematical Rationale*, Chicago, 1907.

WILKINS, John (Bishop of Chester), *Mathematical Magick*, London, 1648.

WILLIAMS, W. M., 'Perpetual Motion on a Large Scale', *Knowledge*, vol. 5, no. 47, 1883–4, p. 29.

WOOD, Edward J., *Curiosities of Clocks and Watches*, London, 1866.

ZIMARA, Mark Antony, *Antrum Magico-Medicum*, Frankfurt, 1625.

INDEX